Fractal Music, Hypercards and More . . .

Mathematical Recreations from SCIENTIFIC AMERICAN Magazine

Fractal Music, Hypercards and More . . .

Mathematical Recreations from SCIENTIFIC AMERICAN Magazine

Martin Gardner

W. H. Freeman and Company
New York

Library of Congress Cataloging-in-Publication Data

Gardner, Martin, 1914–
  Fractal music, hypercards and more : mathematical recreations from Scientific American / by Martin Gardner.

  p. cm.
  Includes index.
  ISBN 0-7167-2188-0. —ISBN 0-7167-2189-9(pbk.)
  1. Mathematical recreations. I. Scientific American. II. Title.
QA95.G26 1991
793.8—dc20 91–17066
CIP

Printed in the United States of America

1 2 3 4 5 6 7 8 9 VB 9 9 8 7 6 5 4 3 2 1

To Douglas Hofstadter

For opening our eyes to the "strange loops" inside our heads, to the deep mysteries of memory, intelligence, and self-awareness, and for the incomparable insights, humor, and wordplay in his writings.

CONTENTS

Preface

This book reprints my Mathematical Games columns from the 1978 and 1979 issues of *Scientific American* Magazine. It is the fourteenth such collection, and I have one more to go before running out of columns. As in previous anthologies, addenda to the chapters update the material with information supplied by faithful readers, and by papers published after the columns were written. In two cases—a column on pi and poetry, and one on minimal sculpture—there was so much to add that I have written new chapters which appear here for the first time.

The book is dedicated to my friend Douglas Hofstadter, whom I first met when he was seeking a publisher for his classic *Gödel, Escher, Bach,* and who later became my successor at *Scientific American.* I had the privilege of reviewing GEB in a column that is reprinted here.

1

White, Brown, and Fractal Music

"For when there are no words [accompanying music] it is very difficult to recognize the meaning of the harmony and rhythm, or to see that any worthy object is imitated by them."

—PLATO, *Laws*, Book II

Plato and Aristotle agreed that in some fashion all the fine arts, including music, "imitate" nature, and from their day until the late 18th century "imitation" was a central concept in western aesthetics. It is obvious how representational painting and sculpture "represent," and how fiction and the stage copy life, but in what sense does music imitate?

By the mid-18th century philosophers and critics were still arguing over exactly how the arts imitate and whether the term is relevant to music. The rhythms of music may be said to imitate such natural rhythms as heartbeats, walking, running, flapping wings, waving fins, water waves, the periodic motions of heavenly bodies and so on, but this does not explain why we enjoy music more than, say, the sound of cicadas or the ticking of clocks. Musical pleasure derives mainly from tone patterns, and nature, though noisy, is singularly devoid of tones. Occasionally wind blows over some object to produce a tone, cats howl, birds warble, bowstrings twang. A Greek legend tells how Hermes invented the lyre: he found a turtle shell with tendons attached to it that produced musical tones when they were plucked.

Above all, human beings sing. Musical instruments may be said to imitate song, but what does singing imitate? A sad, happy, angry or serene song somehow resembles sadness, joy, anger or serenity, but if a melody has no words and invokes no special mood, what does it copy? It is easy to understand Plato's mystification.

There is one exception: the kind of imitation that plays a role in "program music." A lyre is severely limited in the natural sounds it can copy, but such limitations do not apply to symphonic or electronic music. Program music has no difficulty featuring the sounds of thunder, wind, rain, fire, ocean waves and brook murmurings; bird calls (cuckoos and crowing cocks have been particularly popular), frog croaks, the gaits of animals (the thundering hoofbeats in Wagner's *Ride of the Valkyries*), the flights of bumblebees; the rolling of trains, the clang of hammers; the battle sounds of marching soldiers, clashing armies, roaring cannons and exploding bombs. *Slaughter on Tenth Avenue* includes a pistol shot and the wail of a police-car siren. In Bach's *Saint Matthew Passion* we hear the earthquake and the ripping of the temple veil. In the *Alpine Symphony* by Richard Strauss, cowbells are imitated by the shaking of cowbells. Strauss insisted he could tell that a certain female character in Felix Mottl's *Don Juan* had red hair, and he once said that someday music would be able to distinguish the clattering of spoons from that of forks.

Such imitative noises are surely a trivial aspect of music even when it accompanies opera, ballet or the cinema; besides, such sounds play no role whatsoever in "absolute music," music not intended to "mean" anything. A Platonist might argue that abstract music imitates emotions, or beauty, or the divine harmony of God or the gods, but on more mundane levels music is the least imitative of the arts. Even nonobjective paintings resemble certain patterns of nature, but nonobjective music resembles nothing except itself.

Since the turn of the century most critics have agreed that "imitation" has been given so many meanings (almost all are found in Plato) that it has become a useless synonym for "resemblance." When it is made precise with reference to literature or the visual arts, its meaning is obvious and trivial. When it is applied to music, its meaning is too fuzzy to be helpful. In this chapter we take a look at a surprising discovery by Richard F. Voss, a physicist from Minnesota who joined the Thomas J. Watson Research Center of the International Business Machines Corporation after obtaining his Ph.D. at the University of California at Berkeley under the guidance of John Clarke. This work is not likely to restore "imitation" to the lexicon of musical criticism, but it does suggest a curious way in which good music may mirror a subtle statistical property of the world.

The key concepts behind Voss's discovery are what mathematicians and physicists call the spectral density (or power spectrum) of a fluctuating quantity, and its "autocorrelation." These deep notions are technical and hard to understand. Benoît Mandelbrot, who is also at the Watson Research Center, and whose work makes extensive use of spectral densities and autocorrelation functions, has suggested a way of avoiding them here. Let the tape of a sound be played faster or slower than normal. One expects the character of the sound to change considerably. A violin, for example, no longer sounds like a violin. There is a special class of sounds, however, that behave quite differently. If you play a recording of such a sound at a different speed, you have only to adjust the volume to make it sound exactly as before. Mandelbrot calls such sounds "scaling noises."

By far the simplest example of a scaling noise is what in electronics and information theory is called white noise (or "Johnson noise"). To be white is to be colorless. White noise is a colorless hiss that is just as dull whether you play it faster or slower. Its autocorrelation function, which measures how its fluctuations at any moment are related to previous fluctuations, is zero except at the origin, where of course it must be 1. The most commonly encountered white noise is the thermal noise produced by the random motions of electrons through an electrical resistance. It causes most of the static in a radio or amplifier and the "snow" on radar and television screens when there is no input.

With randomizers such as dice or spinners it is easy to generate white noise that can then be used for composing a random "white tune," one with no correlation between any two notes. Our scale will be one octave of seven white keys on a piano: do, re, me, fa, so, la, ti. Fa is our middle frequency. Now construct a spinner such as the one shown at the left in Figure 1. Divide the circle into seven sectors and label them with the notes. It matters not at all what arc lengths are assigned to these sectors; they can be completely arbitrary. On the spinner shown, some order has been imposed by giving fa the longest arc (the highest probability of being chosen) and assigning decreasing probabilities to pairs of notes that are equal distances above and below fa. This has the effect of clustering the tones around fa.

To produce a "white melody" simply spin the spinner as often as you like, recording each chosen note. Since no tone is related in any way to the sequence of notes that precedes it, the result is a totally uncorrelated sequence. If you like, you can divide the circle into more parts and let the spinner select notes that range over the entire piano keyboard, black keys as well as white.

To make your white melody more sophisticated, use another spinner, its circle divided into four parts (with any proportions you like) and labeled 1, 1/2, 1/4 and 1/8 so that you can assign a full, a half, a quarter or an eighth of a beat to each tone. After the composition is completed, tap it out on the piano. The music will sound just like what it is: random music of the dull kind that a two-year-old or a monkey might produce by hitting keys with one finger. Similar white music can be based on random number tables, or the digits in an irrational number.

A more complicated kind of scaling noise is one that is sometimes called Brownian noise because it is characteristic of Brownian motion, the random movements of small particles suspended in a liquid and buffeted by the thermal agitation of molecules. Each particle executes a three-dimensional "random walk," the positions in which form a highly correlated sequence. The particle, so to speak, always "remembers" where it has been.

When tones fluctuate in this fashion, let us follow Voss and call it Brownian music or brown music. We can produce it easily with a spinner and a circle divided into seven parts as before, but now we label the regions, as shown at the right in Figure 1, to represent intervals between successive tones. These step sizes and their probabilities can be whatever we like. On the spinner shown, plus means a step up the scale of

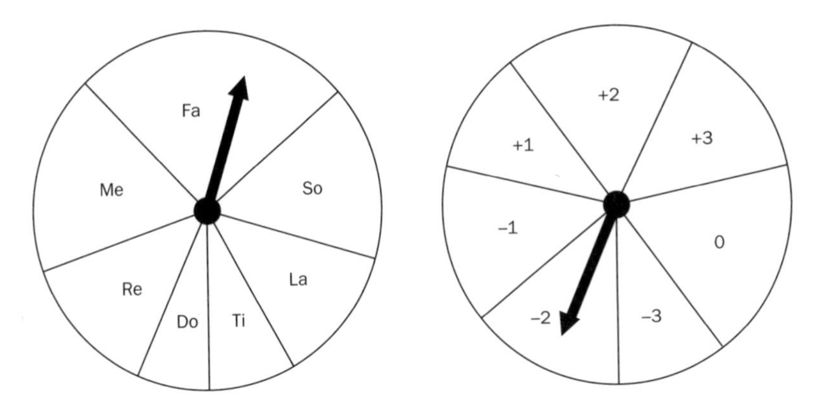

FIGURE 1 Spinners for white music (left) and brown music (right)

one, two or three notes and minus means a step down of the same intervals.

Start the melody on the piano's middle C, then use the spinner to generate a linear random walk up and down the keyboard. The tune will wander here and there, and will eventually wander off the keyboard. If we treat the ends of the keyboard as "absorbing barriers," the tune ends when we encounter one of them. We need not go into the ways in which we can treat the barriers as reflecting barriers, allowing the tune to bounce back, or as elastic barriers. To make the barriers elastic we must add rules so that the farther the tone gets from middle C, the greater is the likelihood it will step back toward C, like a marble wobbling from side to side as it rolls down a curved trough.

As before, we can make our brown music more sophisticated by varying the tone durations. If we like, we can do this in a brown way by using another spinner to give not the duration but the increase or decrease of the duration—another random walk but one along a different street. The result is a tune that sounds quite different from a white tune because it is strongly correlated, but a tune that still has little aesthetic appeal. It simply wanders up and down like a drunk weaving through an alley, never producing anything that resembles good music.

If we want to mediate between the extremes of white and brown, we can do it in two essentially different ways. The way chosen by previous composers of "stochastic music" is to adopt transition rules. These are rules that select each note on the basis of the last three or four. For example, one can analyze Bach's music and determine how often a certain note follows, say, a certain triplet of preceding notes. The random selection of each note is then weighted with probabilities derived from a statistical analysis of all Bach quadruplets. If there are certain transitions that never appear in Bach's music, we add rejection rules to prevent the undesirable transitions. The result is stochastic music that resembles Bach but only superficially. It sounds Bachlike in the short run but random in the long run. Consider the melody over periods of four or five notes and the tones are strongly correlated. Compare a run of five notes with another five-note run later on and you are back to white noise. One run has no correlation with the other. Almost all stochastic music produced so far has been of this sort. It sounds musical if you listen to any small part but random and uninteresting when you try to grasp the pattern as a whole.

Voss's insight was to compromise between white and brown input by selecting a scaling noise exactly halfway between. In spectral terminology it is called $1/f$ noise. (White noise has a spectral density of $1/f^0$, brownian noise a spectral density of $1/f^2$. In "one-over-f" noise the exponent of f is 1 or very close to 1.) Tunes based on $1/f$ noise are moderately correlated, not just over short runs but throughout runs of any size. It turns out that almost every listener agrees that such music is much more pleasing than white or brown music.

In electronics $1/f$ noise is well known but poorly understood. It is sometimes called flicker noise. Mandelbrot, whose book *The Fractal Geometry of Nature* (W. H. Freeman and Company, 1982) has already become a modern classic, was the first to recognize how widespread $1/f$ noise is in nature, outside of physics, and how often one encounters other scaling fluctuations. For example, he discovered that the record of the annual flood levels of the Nile is a $1/f$ fluctuation. He also investigated how the curves that graph such fluctuations are related to "fractals," a term he invented. A scaling fractal can be defined roughly as any geometrical pattern (other than Euclidean lines, planes and surfaces) with the remarkable property that no matter how closely you inspect it, it always looks the same, just as a slowed or speeded scaling noise always sounds the same. Mandelbrot coined the term fractal because he assigns to each of the curves a fractional dimension greater than its topological dimension.

Among the fractals that exhibit strong regularity the best-known are the Peano curves that completely fill a finite region and the beautiful snowflake curve discovered by the Swedish mathematician Helge von Koch in 1904. The Koch snowflake appears in Figure 2 as the boundary of the dark "sea" that surrounds the central motif. (For details on the snowflake's construction, and a discussion of fractals in general, see Chapter 3 of my *Penrose Tiles to Trapdoor Ciphers* (W. H. Freeman, 1989).

FIGURE 2 Mandelbrot's Peano-snowflake as it appeared on the cover of *Scientific American* (April, 1978). The curve was drawn by a program written by Sigmund Handelman and Mark Laff.

The most interesting part of Figure 2 is the fractal curve that forms the central design. It was discovered by Mandelbrot and published for the first time as the cover of *Scientific American*'s April 1978 issue. If you trace the boundary between the black and white regions from the tip of the point of the star at the lower left to the tip of the point of the star at the lower right, you will find this boundary to be a single curve. It is the third stage in the construction of a new Peano curve. At the limit this lovely curve will completely fill a region bounded by the traditional snowflake! Thus Mandelbrot's curve brings together two pathbreaking fractals: the oldest of them all, Giuseppe Peano's 1890 curve, and Koch's later snowflake!

The secret of the curve's construction is the use of line segments of two unequal lengths and oriented in 12 different directions. The curve is much less regular than previous Peano curves and therefore closer to the modeling of natural phenomena, the central theme of Mandelbrot's book. Such natural forms as the gnarled branches of a tree or the shapes of flickering flames can be seen in the pattern.

At the left in Figure 3 is the first step of the construction. A crooked line of nine segments is drawn on and within an equilateral triangle. Four of the segments are then divided

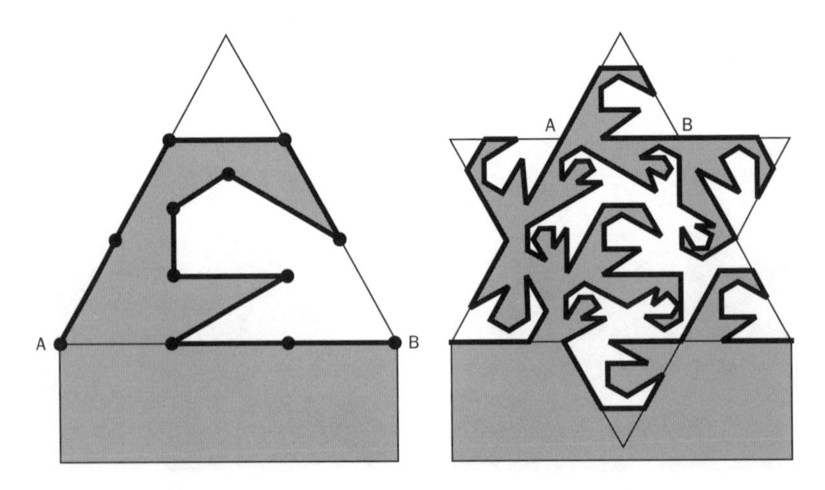

FIGURE 3 The first two steps in constructing Benoît Mandelbrot's Peano-snowflake curve

into two equal parts, creating a line from A to B that consists of 13 long and short segments. The second step replaces each of these 13 segments with a smaller replica of the crooked line. These replicas (necessarily of unequal size) are oriented as is shown inside the star at the right in the illustration. A third repetition of the procedure generates the curve in Figure 2. (It belongs to a family of curves arising from William Gosper's discovery of the "flow-snake," a fractal pictured in Chapter 3 of my above cited book.) When the construction is repeated to infinity, the limit is a Peano curve that totally fills a region bordered by the Koch snowflake. The Peano curve has the usual dimension of 2, but its border, a scaling fractal of infinite length, has (as is explained in Mandelbrot's book) a fractal dimension of log 4/log 3, or 1.2618. . . .

Unlike these striking artificial curves, the fractals that occur in nature—coastlines, rivers, trees, star clustering, clouds and so on—are so irregular that their self-similarity (scaling) must be treated statistically. Consider the profile of the mountain range in Figure 4, reproduced from Mandelbrot's book. This is not a photograph, but a computer-generated mountain scene based on a modified Brownian noise. Any vertical cross section of the topography has a profile that models a random walk. The white patches, representing water or snow in the hollows below a certain altitude, were added to enhance the relief.

The profile at the top of the mountain range is a scaling fractal. This means that if you enlarge any small portion of it, it will have the same statistical character as the line you now see. If it were a true fractal, this property would continue forever as smaller and smaller segments are enlarged, but of course such a curve can neither be drawn nor appear in nature. A coastline, for example, may be self-similar when viewed from a height of several miles down to several feet, but below that the fractal property is lost. Even the Brownian motion of a particle is limited by the size of its microsteps.

Since mountain ranges approximate random walks, one can create "mountain music" by photographing a mountain range and translating its fluctuating heights to tones that fluctuate in time. Villa Lobos actually did this using mountain skylines around Rio de Janeiro. If we view nature statically, frozen in time, we can find thousands of natural curves that can be used in this way to produce stochastic music. Such

FIGURE 4 A modified Brownian landscape generated by a computer program

music is usually too brown, too correlated, however, to be interesting. Like natural white noise, natural brown noise may do well enough, perhaps, for the patterns of abstract art but not so well as a basis for music.

When we analyze the dynamic world, made up of quantities constantly changing in time, we find a wealth of fractal-like fluctuations that have $1/f$ spectral densities. In his book Mandelbrot cites a few: variations in sunspots, the wobbling of the earth's axis, undersea currents, membrane currents in the nervous system of animals, the fluctuating levels of rivers and so on. Uncertainties in time measured by an atomic clock are $1/f$: the error is 10^{-12} regardless of whether one is measuring an error on a second, minute or hour. Scientists tend to overlook $1/f$ noises because there are no good theories to account for them, but there is scarcely an aspect of nature in which they cannot be found.

T. Musha, a physicist at the Tokyo Institute of Technology, discovered that traffic flow past a certain spot on a Japanese expressway exhibited $1/f$ fluctuation. In a more startling experiment, Musha rotated a radar beam emanating from a coastal location to get a maximum variety of landscape on the radar screen. When he rotated the beam once, variations in the distances of all objects scanned by the beam produced a Brownian spectrum. But when he rotated it twice and then subtracted one curve from the other the resulting curve—representing all the changes of the scene—was close to $1/f$.

We are now approaching an understanding of Voss's daring conjecture. The changing landscape of the world (or, to put it another way, the changing content of our total experience) seems to cluster around $1/f$ noise. It is certainly not entirely uncorrelated, like white noise, nor is it as strongly correlated as brown noise. From the cradle to the grave our brain is processing the fluctuating data that comes to it from its sensors. If we measure this noise at the peripheries of the nervous system (under the skin of the fingers), it tends, Mandelbrot says, to be white. The closer one gets to the brain, however, the closer the electrical fluctuations approach $1/f$. The nervous system seems to act like a complex filtering device, screening out irrelevant elements and processing only the patterns of change that are useful for intelligent behavior.

On the canvas of a painting, colors and shapes are static, reflecting the world's static patterns. Is it possible, Mandelbrot asked himself many years ago, that even completely nonobjective art, when it is pleasing, reflects fractal patterns of nature? He is fond of abstract art, and maintains that there is a sharp distinction between such art that has a fractal base and such art that does not, and that the former type is widely considered the more beautiful. Perhaps this is why photographers with a keen sense of aesthetics find it easy to take pictures, particularly photomicrographs, of natural patterns that are almost indistinguishable from abstract expressionist art.

Motion can be added to visual art, of course, in the form of the motion picture, the stage, kinetic art and the dance, but in music we have meaningless, nonrepresentational tones that fluctuate to create a pattern that can be appreciated only over a period of time. Is it possible, Voss asked himself, that the pleasures of music are partly related to scaling noise of $1/f$ spectral density? That is, is this music "imitating" the $1/f$ quality of our flickering experience?

That may or may not be true, but there is no doubt that music of almost every variety does exhibit $1/f$ fluctuations in its changes of pitch as well as in the changing loudness of its tones. Voss found this to be true of classical music, jazz and rock. He suspects it is true of all music. He was therefore not surprised that when he used a $1/f$ flicker noise from a transistor to generate a random tune, it turned out to be more pleasing than tunes based on white and brown noise sources.

Figure 5, supplied by Voss, shows typical patterns of white, $1/f$ and brown when noise values (vertical) are plotted against time (horizontal). These patterns were obtained by a computer program that simulates the generation of the three kinds of sequences by tossing dice. The white noise is based on the sum obtained by repeated tosses of 10 dice. These sums range from 10 to 60, but the probabilities naturally force a clustering around the median. The Brownian noise was generated by tossing a single die and going up one step on the scale if the number was even and down a step if the number was odd.

The $1/f$ noise was also generated by simulating the tossing of 10 dice. Although $1/f$ noise is extremely common in nature, it was assumed until recently that it is unusually cumbersome to simulate $1/f$ noise by randomizers or computers. Previous composers of stochastic music probably did not even know about $1/f$ noise, but if they did, they would have had considerable difficulty generating it. As this article was being prepared Voss was asked if he could devise a simple procedure by which readers could produce their own $1/f$ tunes. He gave some thought to the problem and to his surprise hit on a clever way of simplifying existing $1/f$ computer algorithms that does the trick beautifully.

The method is best explained by considering a sequence of eight notes chosen from a scale of 16 tones. We use three dice of three colors: red, green and blue. Their possible sums range from 3 to 18. Select 16 adjacent notes on a piano, black keys as well as white if you like, and number them 3 through 18.

Write down the first eight numbers, 0 through 7, in binary notation, and assign a die color to each column as is shown in Figure 6. The first note of our tune is obtained by tossing all three dice and picking the tone that corresponds to the

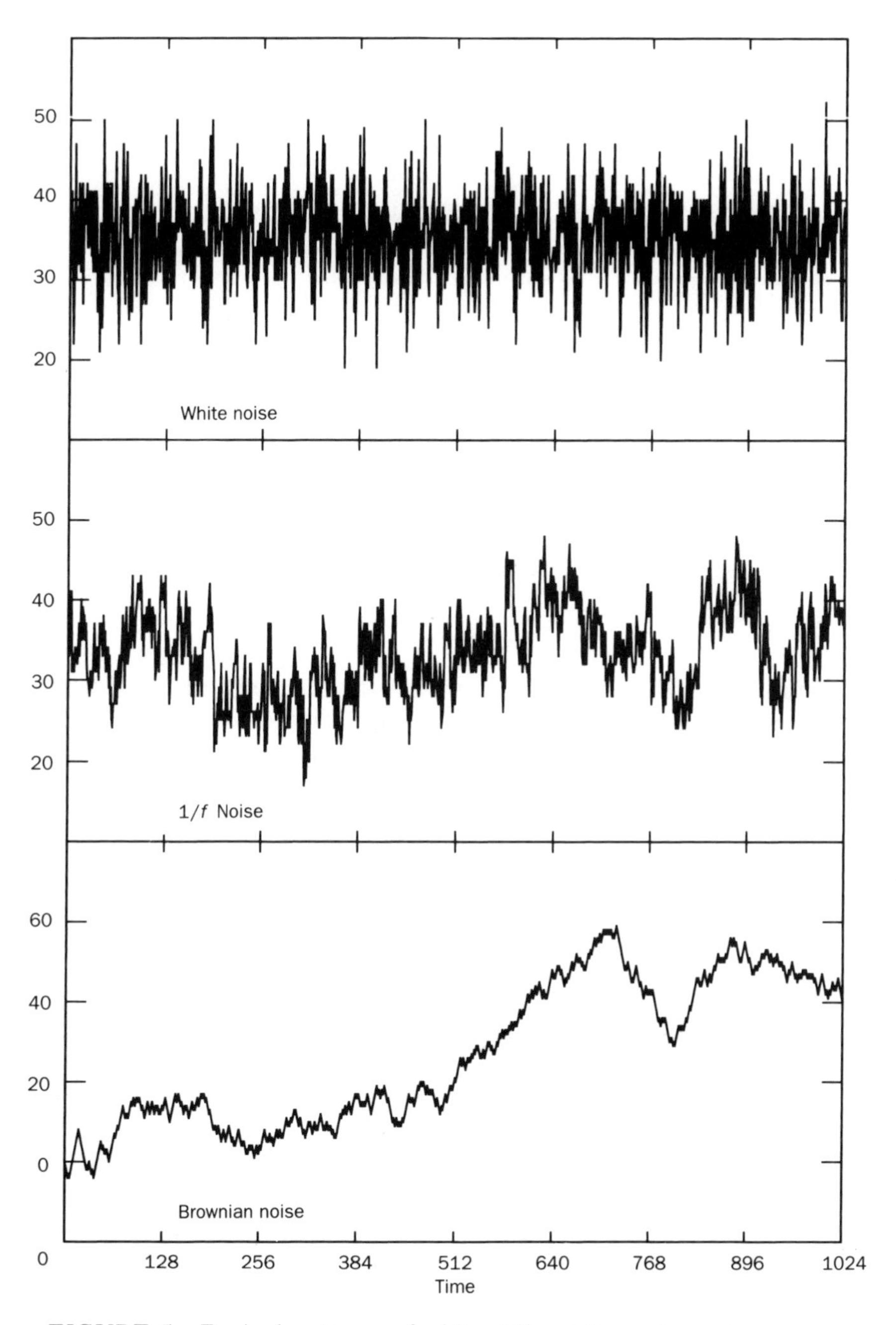

FIGURE 5 Typical patterns of white, 1/*f* and Brownian noise

	Blue	Green	Red
0	0	0	0
1	0	0	1
2	0	1	0
3	0	1	1
4	1	0	0
5	1	0	1
6	1	1	0
7	1	1	1

FIGURE 6 Binary chart for Voss's 1/*f* dice algorithm

sum. Note that in going from 000 to 001 only the red digit changes. Leave the green and blue dice undisturbed, still showing the numbers of the previous toss. Pick up only the red die and toss it. The new sum of all three dice gives the second note of your tune. In the next transition, from 001 to 010, both the red and green digits change. Pick up the red and green dice, leaving the blue one undisturbed, and toss the pair. The sum of all three dice gives the third tone. The fourth note is found by shaking only the red die, the fifth by shaking all three. The procedure, in short, is to shake only those dice that correspond to digit changes.

It is not hard to see how this algorithm produces a sequence halfway between white and brown. The least significant digits, those to the right, change often. The more significant digits, those to the left, are more stable. As a result, dice corresponding to them make a constant contribution to the sum over long periods of time. The resulting sequence is not precisely 1/*f* but is so close to it that it is impossible to distinguish melodies formed in this way from tunes generated by natural 1/*f* noise. Four dice can be used the same way for a 1/*f* sequence of 16 notes chosen from a scale of 21 tones. With 10 dice you can generate a melody of 2^{10}, or 1,024, notes from a scale of 55 tones. Similar algorithms can of course be implemented with generalized dice (octahedrons, dodecahedrons and so on), spinners or even tossed coins.

With the same dice simulation program Voss has supplied three typical melodies based on white, brown, and 1/*f* noise. The computer printouts of the melodies are shown in Figures 7, 8, and 9. In each case Voss varied both the melody and the tone duration with the same kind of noise. Above each tune are shown the noise patterns that were used.

Over a period of two years, tunes of the three kinds were played at various universities and research laboratories, for many hundreds of people. Most listeners found the white music too random, the brown too correlated and the 1/*f* "just about right." Indeed, it takes only a glance at the music itself to see how the 1/*f* property mediates between the two extremes. Voss's earlier 1/*f* music was based on natural 1/*f* noise, usually electronic, even though one of his best compositions derives from the record of the annual flood levels of the Nile. He has made no attempt to impose constant rhythms. When he applied 1/*f* noise to a pentatonic (five-tone) scale and also varied the rhythm with 1/*f* noise, the music strongly resembled Oriental music. He has not tried to improve his 1/*f* music by adding transition or rejection rules. It is his belief that stochastic music with such rules will be greatly improved if the underlying choices are based on 1/*f* noise rather than the white noise so far used.

Note that 1/*f* music is halfway between white and brown in a fractal sense, not in the manner of music that has transition rules added to white music. As we have seen, such music reverts to white when we compare widely separated parts. But 1/*f* music has the fractal self-similarity of a coastline or a mountain range. Analyze the fluctuations on a small scale, from note to note, and it is 1/*f*. The same is true if you break a long tune into 10-note sections and compare them. The tune never forgets where it has been. There is always some correlation with its entire past.

It is commonplace in musical criticism to say that we enjoy good music because it offers a mixture of order and surprise. How could it be otherwise? Surprise would not be surprise if there were not sufficient order for us to anticipate what is likely to come next. If we guess too accurately, say in listening to a tune that is no more than walking up and down the keyboard in one-step intervals, there is no surprise at all. Good music, like a person's life or the pageant of history, is a

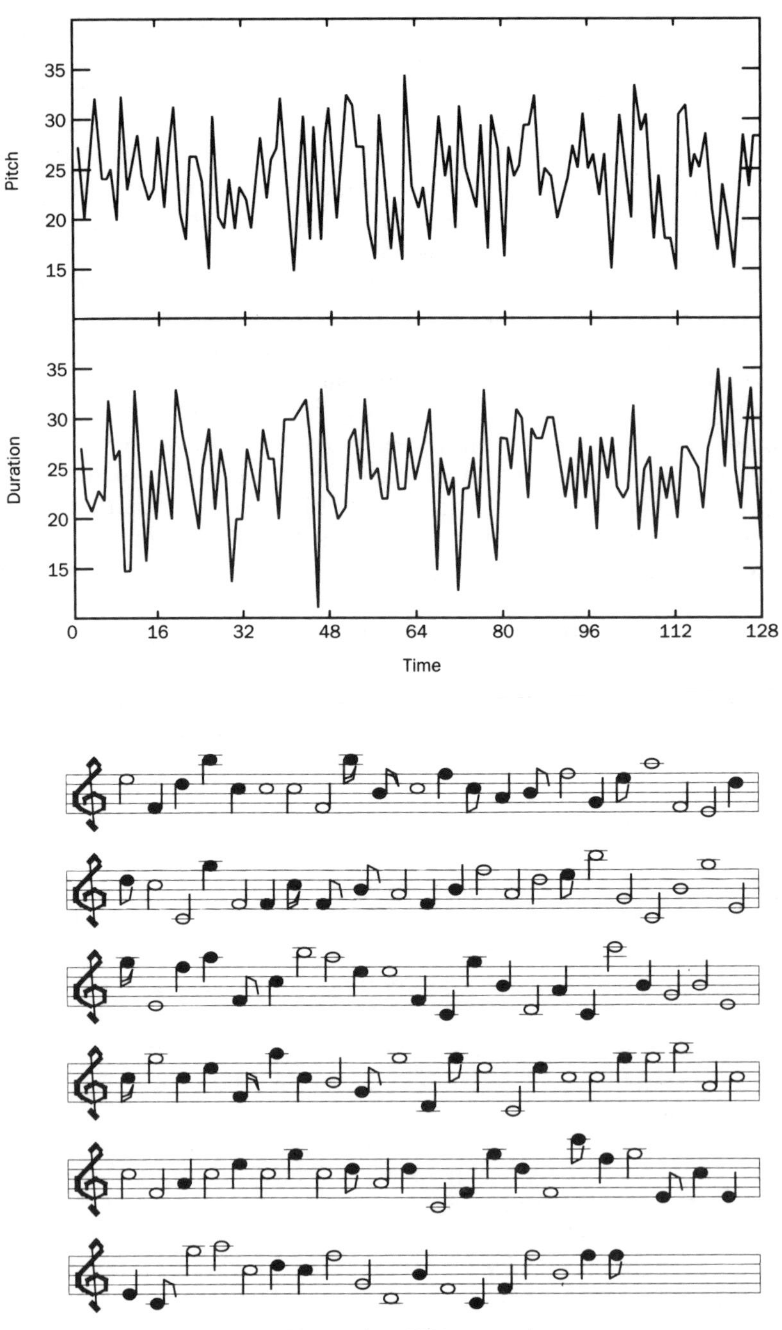

FIGURE 7 White music

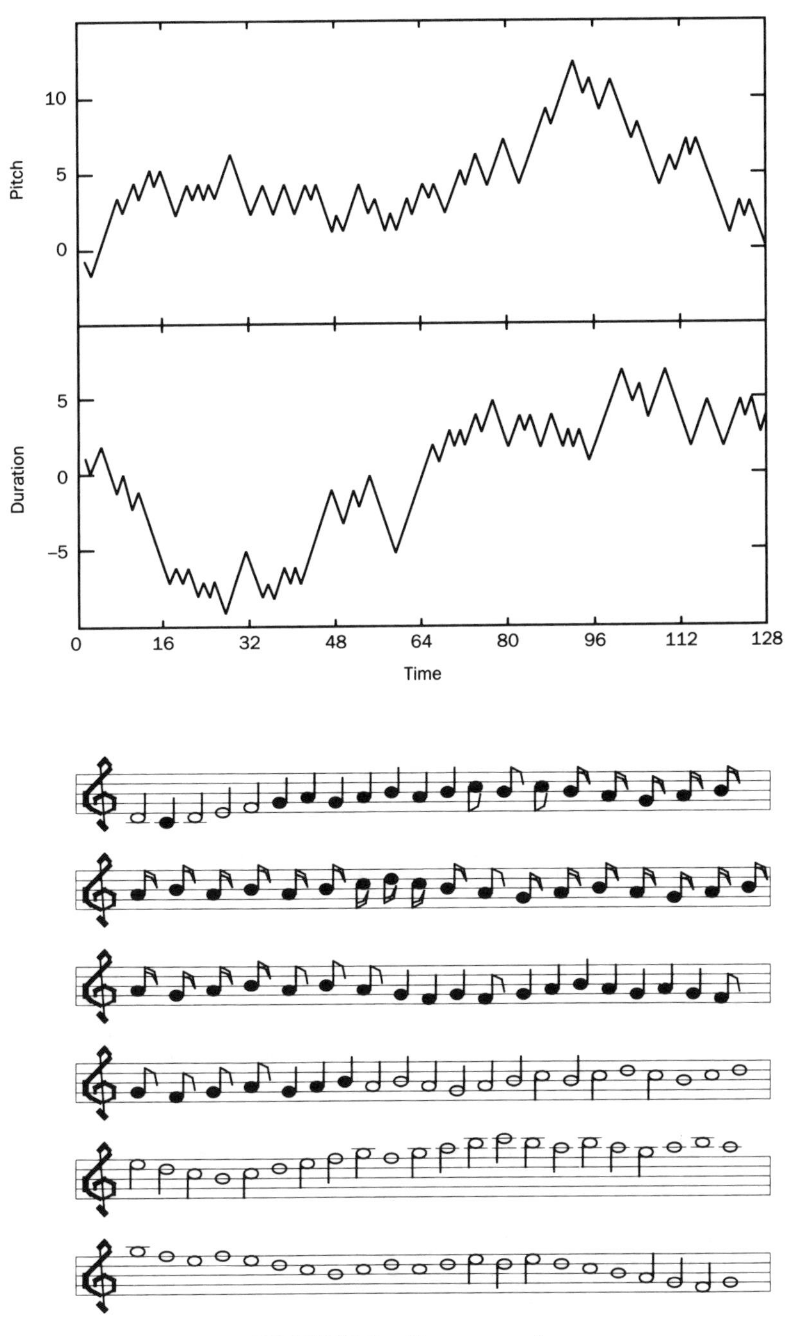

FIGURE 8 Brown music

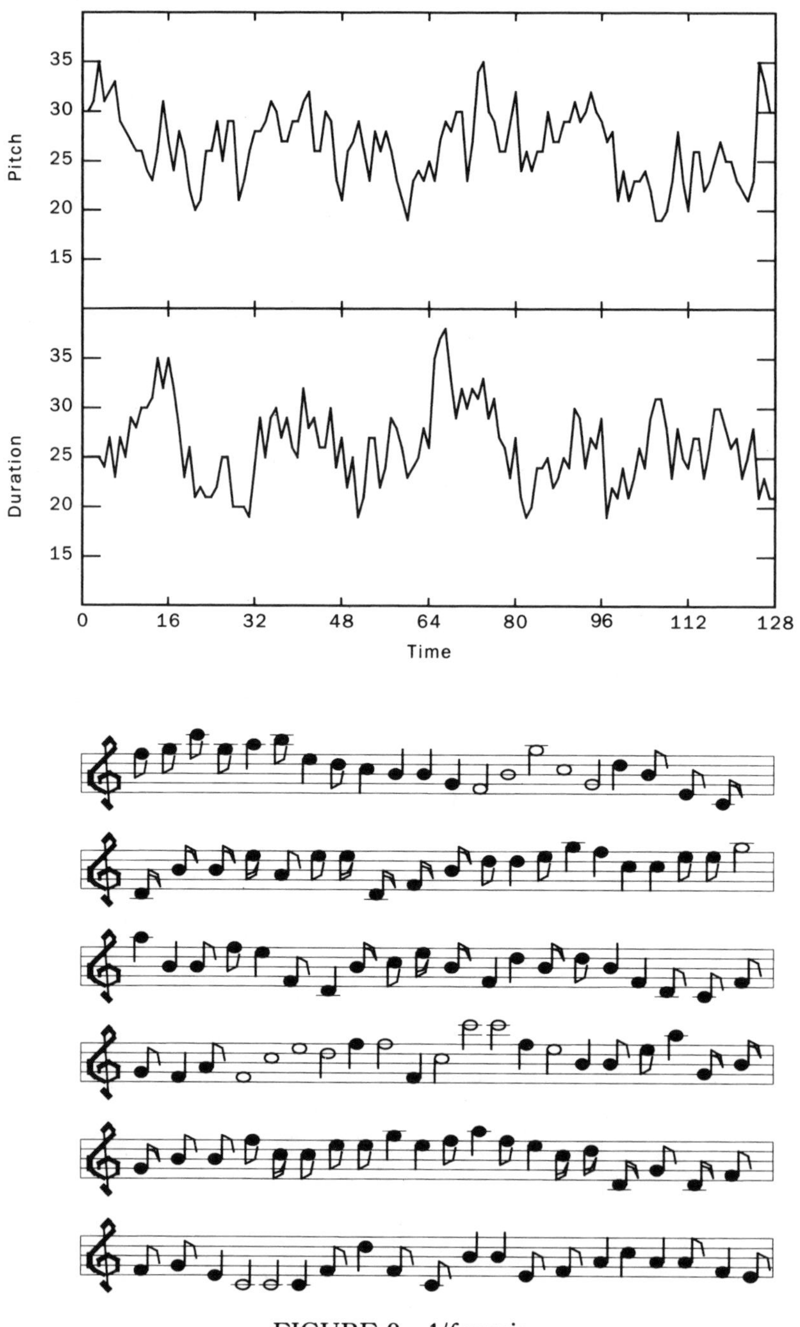

FIGURE 9 1/*f* music

wondrous mixture of expectation and unanticipated turns. There is nothing new about this insight, but what Voss has done is to suggest a mathematical measure for the mixture.

I cannot resist mentioning three curious ways of transforming a melody to a different one with the same $1/f$ spectral density for both tone patterns and durations. One is to write the melody backward, another is to turn it upside down and the third is to do both. These transformations are easily accomplished on a player piano by reversing and/or inverting the paper roll. If a record or tape is played backward, unpleasant effects result from a reversal of the dying-away quality of tones. (Piano music sounds like organ music.) Reversal or inversion naturally destroys the composer's transition patterns, and that is probably what makes the music sound so much worse than it does when it is played normally. Since Voss composed his tunes without regard for short-range transition rules, however, the tunes all sound the same when they are played in either direction.

Canons for two voices were sometimes deliberately written, particularly in the 15th century, so that one melody is the other backward, and composers often reversed short sequences for contrapuntal effects in longer works. Figure 10 shows a famous canon that Mozart wrote as a joke. In this instance the second melody is almost the same as the one you see taken backward and upside down. Thus if the sheet is placed flat on a table, with one singer on one side and the other singer on the other, the singers can read from the same sheet as they harmonize!

No one pretends, of course, that stochastic $1/f$ music, even with added transition and rejection rules, can compete with the music of good composers. We know that certain frequency ratios, such as the three-to-two ratio of a perfect fifth, are more pleasing than others, either when the two tones are played simultaneously or in sequence. But just what composers do when they weave their beautiful patterns of meaningless sounds remains a mystery that even they do not understand.

It is here that Plato and Aristotle seem to disagree. Plato viewed all the fine arts with suspicion. They are, he said (or at least his Socrates said), imitations of imitations. Each time something is copied something is lost. A picture of a bed is not as good as a real bed, and a real bed is not as good as the universal, perfect idea of bedness. Plato was less concerned with the sheer delight of art than with its effects on charac-

FIGURE 10 Mozart's palindromic and invertible canon

ter, and for that reason his *Republic* and *Laws* recommend strong state censorship of all the fine arts.

Aristotle, on the other hand, recognized that the fine arts are of value to a state primarily because they give pleasure, and that this pleasure springs from the fact that artists do much more than make poor copies.

> *They said, "You have a blue guitar,*
> *You do not play things as they are."*
> *The man replied, "Things as they are*
> *Are changed upon the blue guitar."*

Wallace Stevens intended his blue guitar to stand for all the arts, but music, more than any other art and regardless of what imitative aspects it may have, involves the making of something utterly new. You may occasionally encounter natural scenes that remind you of a painting, or episodes in life that make you think of a novel or a play. You will never come on anything in nature that sounds like a symphony. As to whether mathematicians will someday write computer programs that will create good music—even a simple, memorable tune—time alone will tell.

ADDENDUM

Irving Godt, who teaches music history at the Indiana University of Pennsylvania, straightened me out on the so-called Mozart canon with the following letter. It appeared in *Scientific American* (July, 1978):

> A few musical errors slipped past Martin Gardner's critical eye when he took up "Mozart's palindromic and invertible canon" in his report on fractal curves and "one-over-*f*" fluctuations.
>
> Mozart scholars now agree that the canon is almost certainly not by Mozart, even though publishers have issued it under his name. For more than 40 years the compilers of the authoritative Köchel catalogue of Mozart's compositions have relegated it to the appendix of doubtful attributions, where along with three other pieces of a similar character, it bears the catalogue number K. Anh. C 10.16. We have no evidence that the piece goes back any further than the last century.
>
> The piece is not for two singers but for two violins. Singers cannot produce the simultaneous notes of the chords in the second measure (and elsewhere), and the ranges of the parts are quite impractical. To perform the piece the two players begin from opposite ends of the sheet of music and arrive at a result that falls far below the standard of Mozart's authentic canons and other *jeux d'esprit*. The two parts combine for long stretches of parallel octaves, they rarely achieve even the most rudimentary rhythmic or directional independence, and their harmony consists of little more than the most elementary writing in parallel thirds. This little counterfeit is not nearly as interesting as Mr. Gardner's columns.

John G. Fletcher wrote to suggest that because $1/f$ music lies between white and brown music it should be called tan music. The term "pink" has also been suggested, and actually used by some writers. *Fate* magazine (October, 1978) ran a full-page advertisement for an LP record album produced by "Master Wilburn Burchette," of Spring Valley, California, titled *Mind Storm.* The ad calls it "fantastic new deep-hypnotic music that uses a phenomenon known in acoustical science as 'pink sound' to open the mind to thrilling psychic revelations! This astonishing new music acts something like a crystal ball reflecting back the images projected by the mind. . . . Your spirit will soar as this incredible record album carries you to new heights of psychic awareness!"

Frank Greenberg called my attention to some "mountain music" composed by Sergei Prokofiev for Sergei Eisenstein's film *Alexander Nevsky* in 1938. "Eisenstein provided Prokofiev with still shots of individual scenes of the movie as it was being filmed. Prokofiev then took these scenes and used the silhouette of the landscape and human figures as a pattern for the position of the notes on the staff. He then orchestrated around these notes."

BIBLIOGRAPHY

"On Significance in Music." Susanne K. Langer, in *Philosophy in a New Key,* Harvard University Press, 1957.

Noise: Sources, Characterization, Measurement. A. van der Ziel. Prentice-Hall, 1970.

"1/*f* Noise in Music and Speech." Richard F. Voss and John Clarke, in *Nature,* 258, 1975, pages 317–318.

"1/*f* Noise in Music." Richard F. Voss and John Clarke, in *The Journal of the Acoustical Society of America,* 63, 1978, pages 258–263.

"The Noise in Natural Phenomena." Bruce J. West and Michael Shlesinger, in *American Scientist,* 78, 1978, pages 40–45.

"1/*f* Random Tones: Making Music with Fractals." Anthony T. Scarpelli, in *Personal Computing,* 3, 1979, pages 17–27.

"Making Music Fractally." Dietrick E. Thomsen, in *Science News,* 117, 1980, pages 187 ff.

"1/*f* Fluctuations in Biological Systems." T. Musha, in *Sixth International Symposium on Noise in Physical Systems,"* National Bureau of Standards, 1981, pages 143–146.
"Noises: White, Pink, Brown, and Black." Manfred Schroeder, in *Fractals, Chaos, Power Laws,* Chapter 5, W. H. Freeman, 1991.

2

The Tinkly Temple Bells

No! you won't 'eed nothin' else
But them spicy garlic smells
An' the sunshine an' the palm-trees an' the tinkly [Eric]
temple-bells!

—RUDYARD KIPLING, "Mandalay"

Keeping time, time, time,
In a sort of Runic rhyme,
To the tintinnabulation that so
musically wells
From the bells, bells, bells, bells,
Bells, bells, bells—
From the jingling and the tinkling
of the bells.
—EDGAR ALLAN POE, "The Bells"

Imagine that five plates labeled *a, b, c, d* and *e* are in a row on a table. Also on the table are five chessmen: a king, a queen, a bishop, a knight and a rook. In how many ways can the chessmen be arranged on the plates so that one chessman is on each plate? The answer is 5! The exclamation mark is the factorial sign, indicating that the answer is $1 \times 2 \times 3 \times 4 \times 5$, or 120. The problem is combinatorially equivalent to counting the number of ways the letters *a, b, c, d* and *e* can be permuted. In general, for n objects the number of ways is $n!$

Now alter the rules a bit to allow any number of objects, from zero to five, to be on any plate. In how many different ways can the chessmen be placed on the plates? It is obvious that one piece goes on one plate in only one way. Two pieces can go on two plates in four ways, as is shown in Figure 11. If you experiment with three plates and three objects, you will find there are 27 ways. Because 1, 4 and 27 are equal to 1^1, 2^2 and 3^3, one might guess that n objects go on n plates in n^n ways. That is indeed correct. The five chessmen can go on five plates in 5^5, or 3,125, ways.

If there are n objects and k plates, then there are k^n ways to place the objects on the plates according to the altered rules. For example, two objects go on three plates in 3^2, or nine, ways, as is shown at the right in Figure 11. It is easy to see why the formula works. The first object can be placed on the k plates in k different ways. The second object can also go on any of the k plates, so that it too can be placed in k ways. Since there are n objects, it is clear that they can be placed on k plates in $k \times k \times k \times \ . \ . \ . \times k = k^n$ ways.

Consider a more difficult problem. There are the same five chessmen and the same five plates, but now the plates are unlabeled. In other words, the plates are considered to be identical, and so their positions on the table are unimportant. For example, if the king and queen are on one plate and the other three chessmen are on another, it does not matter which plates hold the two groups. All partitions of the set that place the king and queen on one plate and the bishop, knight and rook on another, regardless of which two plates are used, will be considered identical and counted as a single way of placing the chessmen. How can all the ways of placing the five objects be counted?

Once again one object obviously goes on one plate in just one way. Two objects go on two plates in two ways: either both on one plate or one on each plate. This case models many real situations. For example, there are two essentially different ways a husband and wife can occupy unlabeled twin beds: they sleep either in separate beds or in the same bed. There are two ways a policeman can handcuff two prisoners: either

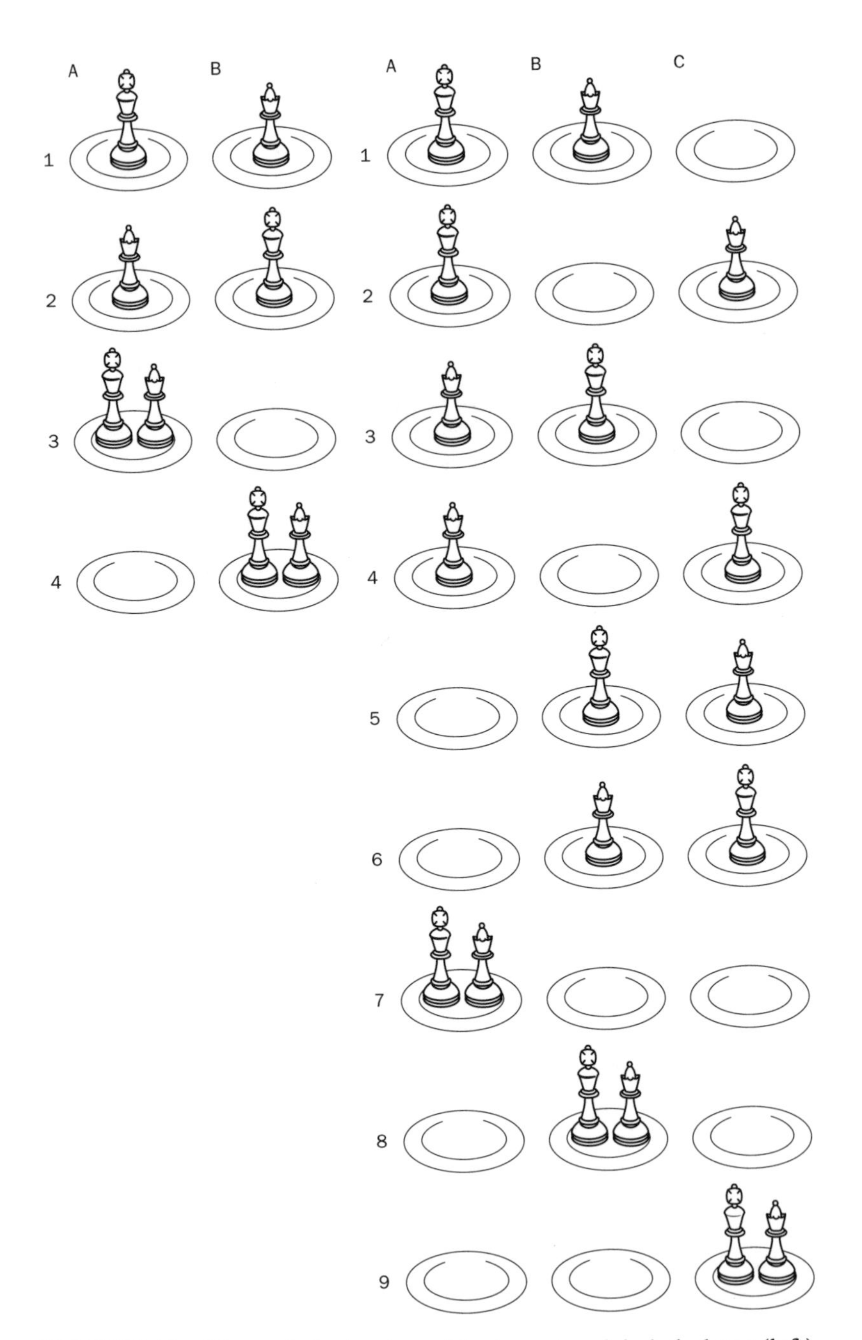

FIGURE 11 Ways of placing two objects on two labeled plates (left) and on three labeled plates (right)

each prisoner can be handcuffed separately or the two can be handcuffed to each other. Figure 12 shows the five ways three objects can be placed on three unlabeled plates. This case models the five ways three people can occupy three unlabeled beds, the five ways three nations can form alliances and so on.

As an experiment you might pause at this point and actually count the ways four objects can be placed on four unlabeled plates. In more technical terms, the problem is to determine the number of ways a set of four distinct elements can be partitioned into nonempty subsets. You will find there are exactly 15 ways. For five objects the number of ways to partition the set jumps to 52. As the number of objects *n* increases, a sequence of numbers is being generated: 1, 2, 5, 15, 52. . . . The numbers in this sequence, which are extremely useful in combinatorial theory, are called Bell numbers in honor of the Scottish-born American mathematician

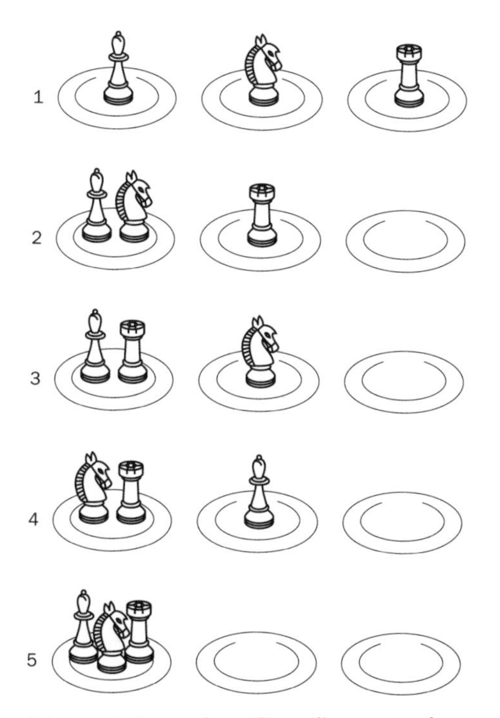

FIGURE 12 Third Bell number ($B_3 = 5$) counts the ways of placing three objects on three unlabeled plates

Eric Temple Bell, who died in 1960. They are closely related to the Catalan numbers, which were the topic of Chapter 20 in my *Time Travel and Other Mathematical Bewilderments* (W. H. Freeman, 1988).

Although Bell numbers were recognized long before Bell wrote about them, he was the first to analyze them in depth and show their importance. In his first paper on the numbers, Bell explained how his interest had been awakened. He had noticed an error in a handbook that gave what is called the Maclaurin expansion for the expression e^{e^x}, where e is the transcendental Euler number and x is any positive integer. The correct expansion is:

$$e\left(1+\frac{x}{1!}+\frac{2x^2}{2!}+\frac{5x^3}{3!}+\frac{15x^4}{4!}\cdots\right)$$

Note that the coefficients for the powers of x are precisely the Bell numbers. (Bell called the numbers exponential numbers, but after the combinatorialist John Riordan began denoting them by B, to honor Bell, they quickly became known as Bell numbers.) From the Maclaurin expansion it is possible to derive what is called Dobinski's formula for the nth Bell number, B_n:

$$B_n=\frac{1}{e}\sum_{k=0}^{\infty}\frac{k^n}{k!}$$

Bell was primarily a number theorist, but he is best known for his classic history *The Development of Mathematics,* his more popularly written *Men of Mathematics* and other books for the general public. Younger mathematicians may be surprised to learn that in the 1920's and 1930's Bell was a prolific writer of science fiction under the pseudonym John Taine. Five of his novels are reprinted in two Dover paperbacks: *Seeds of Life and White Lily* and *The Time Stream.* In 1951 Bell's book *Mathematics, Queen and Servant of Science* was reviewed in a Pasadena Sunday newspaper by John Taine. "The last flap of the jacket," wrote Taine, "says Bell 'is perhaps mathematics' greatest interpreter.' Knowing the author well, the reviewer agrees."

Back to the Bell numbers, which might be called Bells or even Temple Bells. The first 13 Bells are shown in Figure 13, left. By convention B_0 equals 1. As you can see, the numbers grow larger at an exponential rate, or, as Poe has it in "The

Bells," they rise "higher, higher, higher, with a desperate desire." The 100th Bell is a number of 116 digits.

Formulas for the *n*th Bell are complicated and difficult to use in calculating the series, but fortunately there is a simple recursive procedure that cranks them out rapidly. It is best understood by considering the formation of the triangle of numbers shown in Figure 13, top right. (Following a suggestion of correspondent Jeffrey Shallit, I shall call this the Bell triangle.) Start with 1 at the top and 1 below it. Since 1 plus 1 equals 2, place 2 at the end of the second row. Bring the 2 back to start the third row. The sum of 2 and the number above it is 3, and so put 3 to the right of 2. The sum of 3 and the number above it is 5, and so 5 goes to the right of 3. Continue in this manner, observing the following two rules: The last number of each row is the first number of the next row, and all other numbers are obtained by adding the desired number's left neighbor to the number above the neighbor. The sequence of Bell numbers appears on two sides of the triangle. When the triangle is rotated slightly, it becomes a differ-

B_0	1
B_1	1
B_2	2
B_3	5
B_4	15
B_5	52
B_6	203
B_7	877
B_8	4,140
B_9	21,147
B_{10}	115,975
B_{11}	678,570
B_{12}	4,213,597

1
1 2
2 3 5
5 7 10 15
15 20 27 37 52
52 67 87 114 151 203
203 255 322 409 523 674 877
877 . . .
⋮

1 2 5 15 52 203 877 . . .
1 3 10 37 151 674 . . .
2 7 27 114 523 . . .
5 20 87 409 . . .
15 67 322 . . .
52 255 . . .
203 . . .
⋱

FIGURE 13 Bells (left) and two forms of the Bell triangle (right)

ence triangle, as is shown at bottom right of Figure 13. Each number below the top row is the difference of the two numbers above it.

Like the more familiar Pascal triangle, the Bell triangle is rich with interesting properties. In the Bell triangle shown at the top right in the illustration the sums of the horizontal rows are the numbers on the second infinite diagonal. If the sum of a row is added to the Bell number at the end of that row, the number obtained is the next Bell number. If each number is replaced by O for odd or E for even, it is easy to see that every third Bell is even. Hence the ratio of the number of odd Bells to the number of even Bells is 2:1, and the sum of any adjacent triplet of Bells must be even.

Dozens of curious properties of the Bell sequence have been noted, and others are still being discovered. For example, "Touchard's congruence" states: $B_{p+k} \equiv B_k + B_{k+1}$ (modulo p), where p is a prime number. In other words, if the n of B_n is expressed as the sum of a prime p and a number k, and B_n is divided by p, the remainder will equal the remainder obtained when the sum of B_k and B_{k+1} is divided by p. Let k equal zero, and the congruence becomes $B_p \equiv 2$ (modulo p). In other words, every B_n for which n is a prime number has a remainder of 2 when it is divided by that prime. For example, B_{13} is equal to 27,644,437; divide this number by 13 and the remainder is 2.

Bells play an important role in prime-number theory because they count the ways any number with distinct prime factors can be factored. For example, 30 has three different prime factors: 2, 3 and 5. B_3 equals 5. The five ways of factoring 30 are $2 \times 3 \times 5$, 5×6, 3×10, 2×15 and 30. It is not hard to see how this problem is isomorphic with the task of putting three distinct objects on three unlabeled plates. Note that three of the first 10 Bells are prime: 2, 5 and 877. Are there other prime Bells? Is there a largest prime Bell? I do not know the answer to either question.

One of the surprising applications of the Bells is that they count the number of possible rhyme schemes for a stanza of poetry. A one-line stanza has one rhyme scheme, a two-line stanza has two rhyme schemes (the lines either rhyme or do not), a three-line stanza has five *(aaa, aab, aba, abb* and *abc)* and so on. It is said that this "rhyming and the chiming of the bells" (Poe) was first observed by the British mathematician

J. J. Sylvester, but I have not been able to find a reference to it in his little book *The Laws of Verse.* I shall be grateful to any reader who can give me the reference.

H. W. Becker, writing about the Bells in 1941, introduced what he called an "interesting theorem." Call a stanza completely rhymed if every line rhymes with at least one other and incompletely rhymed if at least one line rhymes with no other. The number of possible completely rhymed stanzas for n lines always turns out to equal the number of possible incompletely rhymed stanzas for $n-1$ lines.

Henry W. Gould, a number theorist at West Virginia University, discovered that the Japanese had an attractive way of diagramming rhyme schemes at least as early as A.D. 1000. Figure 14 shows the 52 diagrams for stanzas of five lines. Vertical lines stand for lines of the stanza, and horizontal lines join the lines that rhyme. Gould first described the diagrams in 1976 in his *Research Bibliography of Two Special Number Sequences,* a valuable listing of 175 references on Bell numbers and 445 references on Catalan numbers.

The earliest-known appearance of this method of diagramming is found in *The Tale of Genji,* a famous Japanese novel written by Lady Murasaki, who lived from about A.D. 978 to about 1031. Every chapter except the first and the last in the 54-chapter book is headed by one of the 52 diagrams for stanzas of five lines. The vertical lines are incense sticks, each of which can be any one of five different colors. Horizontal lines join sticks of the same color. The colored diagrams appear in early editions of this Japanese classic but not in English translations. As Joanne Growney observed in her 1970 doctoral dissertation, if all the diagrams with lines that intersect are omitted, the number of remaining diagrams is the fifth Catalan number, 42, and this is true in general for Murasaki diagrams of n lines. Just why Lady Murasaki chose this order for her diagrams, Gould writes, is as unknown as the basis, if indeed there is any, for the ordering of the 64 hexagrams in the *I Ching.*

Quintets are not common in English poetry, but perhaps with diligent searching one could find notable examples of all 52 patterns. For example, Shelley's "To a Skylark" is written in quintets that correspond to the fifth diagram of row five. The fourth diagram of the second row applies to Emily Dickinson's well-know poem:

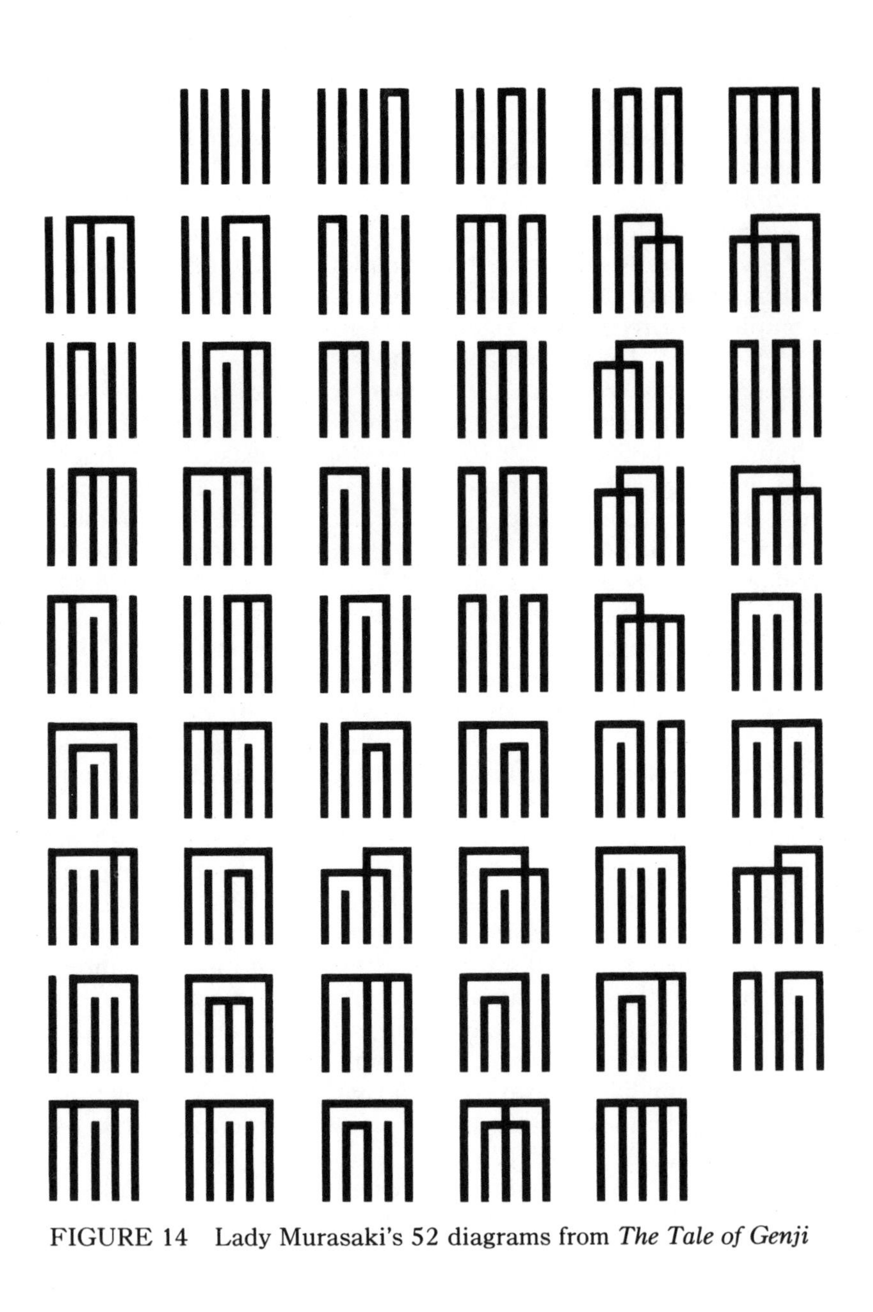

FIGURE 14 Lady Murasaki's 52 diagrams from *The Tale of Genji*

To make a prairie it takes a clover
and one bee,
One clover, and a bee,
And revery.
The revery alone will do,
If bees are few.

Here is a lovely stanza from Alice Meynell's "A Dead Harvest" that corresponds to the pattern of the fourth diagram of row four:

Along the graceless grass of town
They rake the rows of red and brown.
Dead leaves, unlike the rows of hay
Delicate, touched with gold and gray,
Raked long ago and far away.

The limerick is a quintet with a rhyme scheme indicated by the fourth diagram of row six. An unconventional limerick, attributed to W. S. Gilbert, has the scheme of the first diagram:

There was an old man of Dundee,
Who was stung on the arm by a wasp.
When asked "Does it hurt?"
He replied "No it doesn't.
I'm so glad that it wasn't a hornet."

There are several applications of the Bells to graph theory. Consider the following problem. Place six dots in a circle as if to mark the corners of an invisible hexagon and label the dots *a* through *f*. Regard an isolated dot as a degenerate convex polygon of one corner and two dots joined by a straight line as a degenerate convex polygon of two corners. With a pencil connect the dots in any way to form disjoint convex polygons of one, two, three, four, five or six corners. (Disjoint means that no two polygons may have a dot in common.) The lines of the same polygon may not cross one another, because if they did, the polygon would not be convex; lines of distinct polygons may, however, intersect. If you like, you may draw nothing, so that the pattern will consist of six one-corner polygons. Or you may connect all six dots to make a single hexagon. Or you may produce any mixture of polygons provided they are convex and disjoint.

Figure 15 shows four possible patterns. How many different patterns are there? If you have followed the discussion of the Bells, the question should present no difficulty.

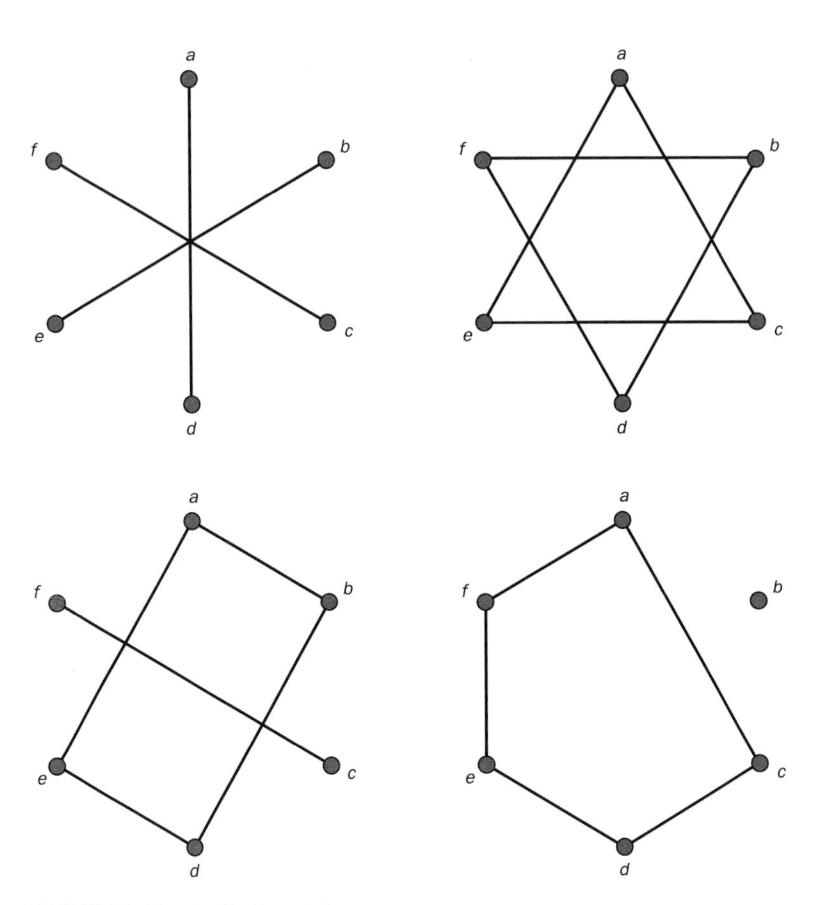

FIGURE 15 A Bell problem

ANSWER

Isolated spots are subsets of one element each, a pair of joined spots is a subset of two elements, three joined spots a subset of three elements, and so on. Each pattern corresponds, therefore, to a way of partitioning a set of elements into disjoint subsets. Since these ways are counted by Bell numbers, the number of patterns for six spots is the sixth Bell number, 203. In general, the number of patterns for n spots is the nth Bell.

ADDENDUM

I raised the question of whether the number of Bell primes is infinite or finite. Many readers pointed out that in addition to the first three Bell primes, B_2, B_3 and B_7, the 13th Bell, 27,644,437, is also prime. A conjecture that all Bell primes have prime subscripts was shot down by Vaughan Pratt of the Laboratory for Computer Science at the Massachusetts Institute of Technology. His fast program, which tested Bells through B_{161} (a number of more than 200 digits), found the next two Bell primes, B_{42} and B_{55}. Pratt conjectures that there are infinitely many Bell primes but that he will learn of at most one new one in his lifetime. B_{42} has 38 digits; B_{55} has 54.

Sin Hitotumatu provided more details on the Lady Murasaki diagrams. In 1600 Japanese noblemen and ladies played a game called Genji-ko or Monko. An umpire randomly selected sticks of incense from a supply that contained five different kinds. Players sniffed the burning sticks and tried to guess which were the same and which were different. The 52 possible selections were diagrammed by the players as explained earlier.

In the early 17th century, Japanese mathematicians assigned a mnemonic name to each diagram, using the names of the 52 chapters (between the first and the last) of Lady Murasaki's novel *Tale of Genji*. It is not known whether this assignment of diagrams to chapters was random or was based on some pattern or perhaps a fancied correlation with the characters and events of each chapter. In the late 19th century, printed editions of *Tale of Genji* began to carry the Murasaki diagrams as chapter headings.

Andrew Lenard of Indiana University reported success in proving the following curious property of Bell numbers. (The property had been noticed but not established.) The first $2n$ Bells can be arranged in a square matrix for $n=3$:

$$\begin{matrix} 1 & 1 & 2 & 5 \\ 1 & 2 & 5 & 15 \\ 2 & 5 & 15 & 52 \\ 5 & 15 & 52 & 203 \end{matrix}$$

What is the determinant of such a matrix? It is given by an astonishingly simple formula: $(1!)(2!)(3!)\ .\ .\ .\ (n!)$. In this case 203 is the sixth Bell; $n=3$, and the formula gives $1 \times 2 \times 6 = 12$ as the determinant's value.

Christian Radoux, of the Universite de l'Etat, in Belgium, had guessed the correctness of the formula in a note in the *Notices of the American Mathematical Society,* 25, 1978, p. 197. He wrote that he later proved the result, and also generalized it in a paper scheduled to appear in a French journal.

Antoni Kanczewski pointed out that the stanzas of Robert Frost's "The Road Not Taken" have the rhyme scheme of the last diagram in the second row of Lady Murasaki's figures. Here is the familiar last stanza:

I shall be telling this with a sigh
Somewhere ages and ages hence:
Two roads diverged in a wood and I—
I took the one less travelled by,
And that has made all the difference.

Philip Doty reminded me that the rhymeless Limerick credited to W. S. Gilbert was given in numerous variations in letters to the *London Times Literary Supplement,* during the spring and summer of 1978. Anthology references are cited.

The most eccentric rhyme scheme known to me was used by Dylan Thomas for the prologue to his *Collected Poems.* (I came across this in Bob Vannicombe's "Quiz—Mathematics in Literature," *Journal of Recreational Mathematics,* 10, 1977–78, pp. 267–269.) The scheme can be expressed by numbers 1 through 51, followed by the same numbers in reverse order. Thomas was reportedly upset by the fact that few readers noticed the pattern.

Alan Watton, Sr., in a letter in *Scientific American* (July, 1978) disclosed that Eric Temple Bell's first published article on mathematics was in *Scientific American* in 1916. The paper won honorable mention in a contest for the best explanation of relativity theory written for persons of average intelligence.

Ethan Bolker of the University of Massachusetts and David Robbins of Hamilton College discovered a surprising application of Bell numbers to card shuffling. Given a packet of n playing cards, we define a shuffle as follows. The first (top) card is placed at any position in the packet from 1 through n. (If it is placed at 1, of course, it stays where it is.) The card now on top is placed at any position 1 through n. The procedure is repeated n times.

For a packet of n cards there are n^n possible shuffles. How many restore the packet to its original order? The answer is the nth Bell number.

We can describe each shuffle by a sequence of numbers that gives the positions to which each top card is shifted. For example, if in the third move the top card goes to second from the bottom in a packet of 10, the third number in the sequence is 9. The pattern of a shuffle is uniquely defined by this chain of position numbers.

When n is 1, the only possible shuffle, 1, is trivially a Bell shuffle, that is, a shuffle that restores the pack to its original order. When n is 2, there are two Bell shuffles: 11 and 22. When n is 3, there are 3^3, or 27, possible shuffles, of which five (111, 122, 212, 221 and 333) are Bell shuffles. When n is 4, there are 15 shuffles that restore order, and so on. As n increases, the Bells are generated.

Other shuffling procedures are similarly related to Bell numbers. For instance, the shuffle described above can be modified so that instead of the top card being shifted on every move it is shifted only on the first move. The second card of the new arrangement is shifted on the second move, the third card on the third move and so on. Once again the nth Bell number counts the shuffles that restore the original order of n cards. For example, the five Bell shuffles for three cards are 123, 132, 213, 231 and 321.

Still another shuffle can be described by assuming you have a packet of playing cards with an ace on top, a deuce second, a three third, and so on. Put the ace anywhere, then put the deuce anywhere, the three anywhere, and continue to the nth value. For three cards, the five Bell shuffles that restore the initial order are 123, 133, 223, 232 and 333.

Bolker and Robbins found an ingenious way of establishing a one-to-one correspondence between the Bell shuffles and the set of partitions counted by the Bells. Their paper on these nonrandom shuffles, and others related to Catalan numbers, is given in the bibliography.

BIBLIOGRAPHY

"Exponential Numbers." Eric T. Bell, in *The American Mathematical Monthly,* 41, 1934, pages 411–419.

"Solution to Problem E 461." H. W. Becker, in *The American Mathematical Monthly,* 48, 1941, pages 701–703.

"The Number of Partitions of a Set." Gian-Carlo Rota, in *The American Mathematical Monthly,* 71, 1964, pages 498–504.

"The Theory of Partitions." George E. Andrews, in *Encyclopedia of Mathematics and Its Applications:* Vol. 2, edited by Gian-Carlo Rota, Addison Wesley, 1976.

"A Wisp of Smoke: Scent and Character in *The Tale of Genji.*" Aileen Gatten, in *Monumenta Nipponica,* 32, 1977, pages 35–48.

The Tale of Genji. Murasaki Shikubu. Knopf, 1978.

"A Budget of Rhyme Scheme Counts." John Riordan, in *Annals of the New York Academy of Sciences,* 319, 1979, pages 445–465.

"The Bias of Three Pseudo-random Shuffles." D. P. Robbins and E. D. Bolker, in *Aequationes Mathematicae,* 22, 1981, pages 268–292.

Mathematical Zoo

There has never been a zoo designed to display animals with features of special interest to recreational mathematicians, yet such a zoo could be both entertaining and instructive. It would be divided, as I visualize it, into two main wings, one for live animals, the other for pictures, replicas and animated cartoons of imaginary creatures. Patrons of the "mathzoo" would be kept informed of new acquisitions by a newsletter called ZOONOOZ (with the permission of the Zoological Society of San Diego, which issues a periodical of that name), a title that is both palindromic and the same upside down.

A room of the live-animal wing would contain microscopes through which one could observe organisms too tiny to be seen otherwise. Consider the astonishing geometrical symmetries of radiolaria, the one-celled organisms that flourish in the sea. Their intricate silica skeletons are the nearest counterparts in the biological world to the patterns of snow crystals. In his *Monograph of the Challenger Radiolaria,* the German biologist Ernst Haeckel described thousands of radiolaria species that he discovered on the *Challenger* expedition of 1872–76. The book contains 140 plates of drawings that have never been excelled in displaying the geometric details of these intricate, beautiful forms.

Figure 16 from Haeckel's book, is of special interest to mathematicians. The first radiolarian is basically spherical, but its six clawlike extensions mark the corners of a regular octahedron. The second skeleton has the same solid at its center. The third is a regular icosahedron of 20 faces. The fifth is the 12-sided dodecahedron. Other plates in Haeckel's book show radiolaria that approximate cubical and tetrahedral forms.

It is well known that there are just five Platonic solids, three of which have faces that are equilateral triangles. Not so widely known is that there are an infinite number of semiregular solids also with sides that are equilateral triangles. They are called "deltahedra" because their faces resemble the Greek letter delta. Only eight deltahedra are convex: those with 4, 6, 8, 10, 12, 14, 16 and 20 faces. The missing 18-sided convex deltahedron is mysterious. One can almost prove it should exist, and it is not so easy to show why it cannot. It is hard to believe, but the proof that there are only eight convex deltahedra was not known until B. L. van der Waerden and Hans Freudenthal published it in 1947. If concavity is allowed, a deltahedron can have any number of faces of eight or greater.

The four-faced deltahedron is the regular tetrahedron, the simplest of the Platonic solids. The six-faced deltahedron consists of two tetrahedra sharing one face. Note the fourth radiolarian in Haeckel's picture. It is a 10-faced deltahedron, or rather one that is inflated slightly toward a sphere. It may surprise you to learn that there are two topologically distinct eight-sided deltahedra. One is the familiar regular octahedron. Can you construct a model of the other one (it is not convex)?

Surfaces of radiolaria are often covered with what seems to be a network of regular hexagons. The regularity is particularly striking in *Aulonia hexagona,* shown in Figure 17. Such networks are called "regular maps" if each cell has the same number of edges and each vertex has the same number of edges joined to it. Imagine a regular tetrahedron, octahedron or icosahedron inflated like a balloon but preserving its edges as lines on the resulting sphere. The tetrahedron will form a regular map of triangles with three edges at each vertex, the octahedron a map of triangles with four edges at each vertex, and the icosahedron a map of triangles with five edges at each

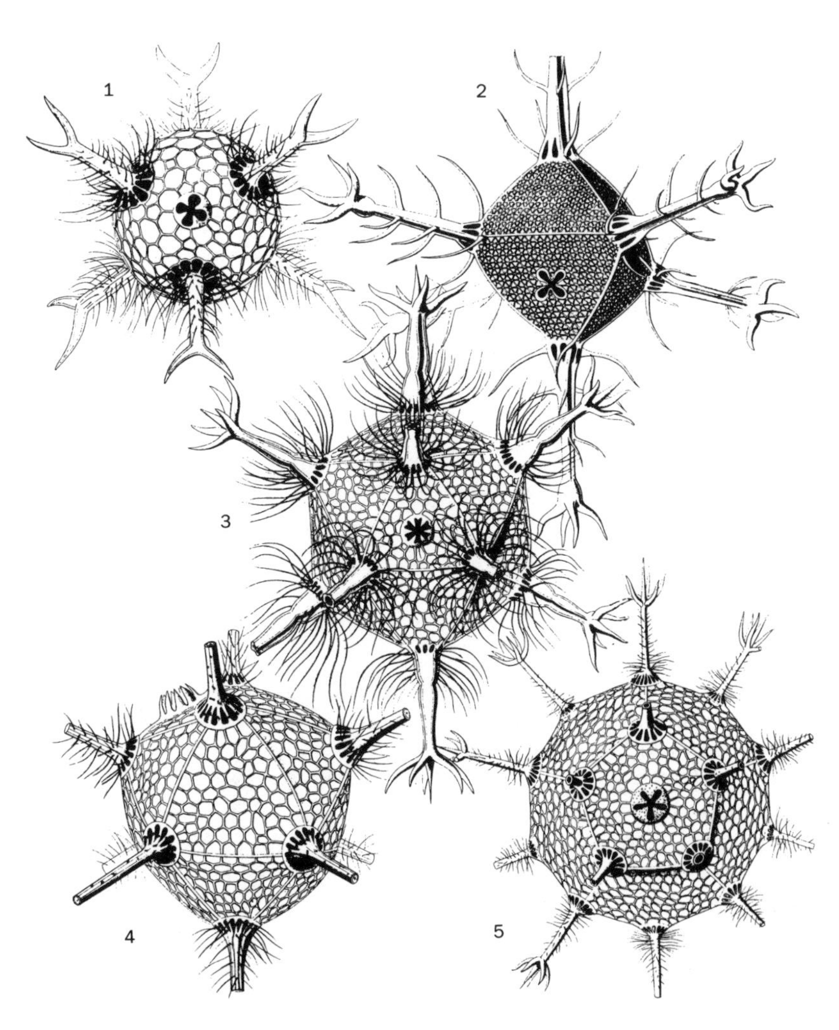

FIGURE 16 Radiolaria skeletons in Ernst Haeckel's *Monograph of the Challenger Radiolaria*

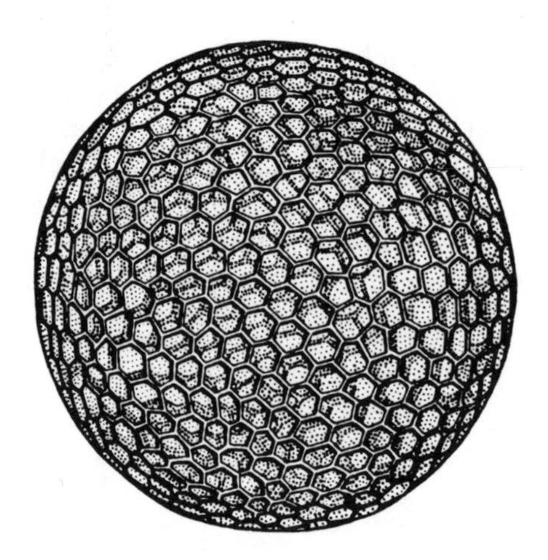

FIGURE 17 The radiolarian *Aulonia hexagona*

vertex. Inflating a cube produces a regular map of four-sided cells with three edges at each vertex. Inflating a dodecahedron produces a regular map of pentagons with three edges at each vertex.

Aulonia hexagona raises an interesting question. Is it possible to cover a sphere with a regular map of hexagons, three edges at each vertex? Only the topological properties of the map concern us. The hexagons need not be regular or even convex. They may have any size or shape, and their edges may twist and curve any way you like provided they do not intersect themselves or one another and provided three of them meet at each vertex.

The answer is no, and it is not hard to prove impossibility with a famous formula that Leonhard Euler discovered for the skeletons of all simply connected (no "holes") polyhedra. The formula is $F+C-E=2$, where the letters stand for faces, corners and edges. Since all such polyhedra can be inflated to spheres, the formula applies also to maps on the sphere. In Chapter 13 of *Enjoyment of Mathematics,* by Hans Rademacher and Otto Toeplitz, you will find it explained how Euler's formula can be used in proving that no more than five regular maps can be drawn on a sphere and that therefore no more than five regular convex solids can exist. As a second problem, can you use Euler's formula to show that a regular map of hexagons is impossible on a sphere?

D'Arcy Wentworth Thompson, whose classic work *On Growth and Form* contains an excellent section on radiolaria, liked to tell about a biologist who claimed to have seen a spherical radiolarian covered with a perfect map of hexagons. But, said Thompson, Euler had proved this impossible. "That," replied the biologist, "proves the superiority of God over mathematics."

"Euler's proof happened to be correct," writes Warren S. McCulloch in an essay where I found this anecdote, "and the observations inaccurate. Had both been right, far from proving God's superiority to logic, they would have impugned his wit by catching him in a contradiction." If you look carefully at the picture of *Aulonia hexagona* you will see cells with more or fewer than six sides.

Under electron microscopes in our zoo's micro room would be the many viruses that recently have been found to crystallize into macromolecules shaped like regular icosahedra: the measles virus, the herpes, the triola iridescent and many others (see R. W. Horne's article cited in the bibliography). Viruses may also have dodecahedral shapes, but as far as I know this remains unsettled. Another recent discovery is that some viruses, such as the one that causes mumps, are helical. It had formerly been thought that helical structures were restricted to plants and to parts of animals: hair, the umbilical cord, the cochlea of the human ear, the DNA molecule and so on. A section of our zoo would feature such spectacular helical structures as molluscan seashells, the twisted horns of certain sheep, goats, antelopes and other mammals, and such curiosities as "devil's corkscrews"—the huge fossil burrows of extinct beavers (see Chapter 1, "The Helix," of my *Sixth Book of Mathematical Games from Scientific American*).

In the macro world of fishes, reptiles, birds, insects, mammals and human beings the most striking geometrical aspect of the body is its overall bilateral symmetry. It is easy to understand why this symmetry evolved. On the earth surface gravity creates a marked difference between up and down, and locomotion creates a marked difference between front and back. But for any moving, upright creature the left and right sides of its surroundings—in the sea, on the land or in the air—are fundamentally the same. Because an animal needs to see, hear, smell and manipulate the world equally well on both sides, there is an obvious survival value in having nearly identical left and right sides.

Animals with bilateral symmetry are of no interest for our mathzoo—you can see them at any zoo—but it would be amusing to assemble an exhibit of the most outrageous violations of bilateral symmetry. For example, an aviary would feature the crossbill, a small red bird in the finch family that has its upper and lower beaks crossed in either of the two mirror-image ways. The bird uses its crossed bill for prying open evergreen cones in the same way a cook uses a plierlike device to pry off the lid of a jar or can. A medieval legend has it that the bill became twisted as the bird was trying vainly to pull the nails from the cross when Jesus was crucified; in the effort the bird's plumage became stained with blood. In the same aviary would be some wry-billed plovers from New Zealand. The entire bill of this funny bird is twisted to the right. The bill is used for turning over stones to find food. As you would expect, foraging wry-billed plovers search mainly on the right.

An aquarium in our mathzoo would exhibit similar instances of preposterous asymmetry among marine life: the male fiddler crab, for example, with its enormous left (or right) claw. Flatfish are even more grotesque examples. The young are bilaterally symmetric, but as they grow older one eye slowly migrates over the top of the head to the other side. The poor fish, looking like a face by Picasso, sinks to the bottom, where it lies in the ooze on its eyeless side. The eyes on top turn independently so that they can look in different directions at the same time.

Another tank would contain specimens of the hagfish. This absurd fish looks like an eel, has four hearts, teeth on its tongue and reproduces by a technique that is still a mystery. When its single nostril is clogged, it sneezes. The hagfish is in our zoo because of its amazing ability to tie itself into an overhand knot of either handedness. By sliding the knot from tail to head it scrapes slime from its body. The knot trick is also used for getting leverage when the hagfish tears food from a large dead fish and also for escaping a predator's grasp (see David Jensen's article listed in the bibliography).

Knots are, of course, studied by mathematicians as a branch of topology. Another exhibit in our aquarium would be beakers filled with *Leucothrix mucor,* a marine bacterium shaped like a long filament. A magnifying glass in front of each beaker would help visitors see the flimsy filaments. They reproduce by tying themselves into knots—overhands, figure-eights, even more complicated knots—that get tighter and tighter until they pinch the filament into two or more parts (see Thomas D. Brock's paper listed in the bibliography). Do higher animals ever tie parts of themselves into knots? Fold your arms and think about it.

The most popular of our aquarium exhibits would probably be a tank containing specimens of *Anableps,* a small (eight-inch) Central American carp sometimes called the stargazer. It looks as if it has four eyes. Each of its two bulging eyes is divided into upper and lower parts by an opaque band. There is one lens but separate corneas and irises. This little BEM (bug-eyed monster) swims with the band at water level. The two upper "eyes" see above water while the two lower ones see below. The *Anableps* is in our zoo because of its asymmetric sex life. The young are born alive, which means that the male must fertilize the eggs inside the female. The female opening is on either the left side or the right. The male organ also is either on the left or the right. This makes it impossible for two fish of the same handedness to mate. Fortunately both males and females are equally left- or right-sexed, and so the species is in no danger of extinction.

In a larger tank one would hope to see some narwhals, although until now they have not survived in captivity. This small whale, from north-polar seas, has been called the sea unicorn because the male has a single "horn" that projects straight forward from its upper jaw and is about half the whale's body length. Both sexes are born with two small side-by-side teeth. The teeth stay small on the female, but the male's left tooth grows into an ivory tusk, straight as a javelin and seven to 10 feet long. This ridiculous tooth, the longest in the world, has a helical groove that spirals around it like a stripe on a barber pole. Nobody knows what function the tusk serves. It is not used for stabbing enemies or punching holes in ice, but during the mating season narwhals have been seen fencing with each other, so that its main purpose may be a role in sexual ritual (see John Tyler Bonner's article in the bibliography). Incidentally, the narwhal is also unusual in having a name starting with the letter *n.* It is easy to think of mammalian names beginning with any letter of the alphabet except *n.*

Among snakes, species that sidewind across the desert sands are mathematically interesting because of their highly asymmetric tracks: sets of parallel line segments that slant either right or left at angles of about 60 degrees from the line of travel. Many species of snakes are capable of sidewinding, notably the sidewinder itself, a small rattlesnake of Mexico and the U.S. Southwest, and the African desert viper. Exactly how sidewinding works is rather complicated, but you will find it clearly explained in Carl Gans's article.

The insect room of our mathzoo would certainly display the nests of bees and social wasps. They exhibit a hexagonal tessellation even more regular than the surfaces of radiolaria. A large literature, going back to ancient Greece and still growing, attempts to explain the factors that play a role in producing this pattern. D'Arcy Thompson, in his book cited earlier, has a good summary of this literature. In times before Darwin bees were usually regarded as being endowed by the Creator with the ability to design nests so that the cells use the least amount of wax to hold a maximum amount of honey. Even Darwin marveled at the bee's ability to construct a honeycomb, calling that ability "the most wonderful of known instincts," and "absolutely perfect in economizing labor and wax."

Actual honeycombs are not as perfect as early writers implied, and there are ways of tessellating space with polyhedral cells that allow an even greater economy of wax. Moreover, it seems likely that the honeycomb pattern is less the result of evolution finding a way to conserve wax than an accidental product of how bees use their bodies and the way they form dense clusters when they work. Surface tension in the semiliquid wax may also play a role. The matter is still far from settled. The best discussion I know is a paper by the Hungarian mathematician L. Fejes Tóth.

No actual animal propels itself across the ground by rolling like a disk or a sphere, but our insect room would be incomplete without an exhibit of a remarkable insect that transports its food by rolling near-perfect spheres. I refer to the dung beetle, the sacred scarab of ancient Egypt. These sometimes beautiful insects (in the Tropics they have bright metallic colors) use their flat, sharp-edged heads as shovels to dig a supply of fresh ordure that their legs then fashion into spheres. By pushing with its hind legs and walking backward the dung beetle will roll the little ball to its burrow where it will be consumed as food. No one has described the process with more literary skill and humor than the French entomologist, Jean Henri Fabre, in his essay on "The Sacred Beetle."

Our zoo's imaginary wing would lack the excitement of living creatures but would make up for it in wild fantasy. In

Flaubert's *Temptation of St. Anthony,* for example, there is a beast called the Nasnas that is half of an animal bisected by its plane of symmetry. Jorge Luis Borges, in his delightful *Book of Imaginary Beings,* refers to an earlier invention of such a creature by the Arabs. L. Frank Baum's fantasy, *Dot and Tot of Merryland,* tells of a valley inhabited by wind-up animals. The toys are kept wound by a Mr. Split, whose left half is bright red and right half white. He can unhook his two sides, each of which hops about on one leg so that he gets twice as much winding done. Conversing with a half of Mr. Split is difficult because Mr. Left Split speaks only the left halves of words and Mr. Right Split only the right halves.

A variety of mythical "palindromic" beasts violate front and back asymmetry by having identical ends. Borges writes of the fabled *amphisbaena* (from the Greek for "go both ways"), a snake with a head at each end. Dante puts the snake in the seventh circle of Hell, and in Milton's *Paradise Lost* some of Satan's devils are turned into *amphisbaenas.* Alexander Pope writes in his *Dunciad:*

Thus Amphisbaena (I have read)
At either end assails;
None knows which leads,
or which is led,
For both Heads are but Tails.

The fable is not without foundation. There are actual snakes called amphisbaenas that crawl both ways and have such tiny eyes that it is hard to distinguish one end from the other. If a flatworm's head is cut off, another grows at the base of the severed head, so palindromic animals actually can exist. In Baum's *John Dough and the Cherub* one meets Duo, a dog with a head and forelegs at both ends (see Figure 18). The animal anticipates the Pushmi-Pullyu (it has a two-horned head at each end) that flourishes in the African jungle of Hugh Lofting's Dr. Dolittle books.

Rectangular parallelepipeds are never the parts of real animals, but in Baum's *Patchwork Girl of Oz* there is a block-headed, thick-skinned, dark blue creature called the Woozy (see Figure 19). The animal's head, body, legs and tail are shaped like blocks. It is friendly as long as no one says "Krizzle-kroo." This makes the Woozy so angry that its eyes dart fire. Nobody, least of all the Woozy, knows what Krizzle-kroo

FIGURE 18 Duo, a palindromic dog

FIGURE 19 L. Frank Baum's Woozy

means, and that is what makes it so furious. Borges reminds us of the Gillygoo, a bird in the Paul Bunyan mythology, that nests on steep slopes and lays cubical eggs that will not roll down and break. Minnesota lumberjacks hard-boil them and use them for dice. In Stanley G. Weinbaum's story, "A Martian Odyssey," a species of nondescript animals on Mars excrete silica bricks that they use for building pyramidal dwellings.

Baum also imagined spherical creatures. The Roly-Rogues, in *Queen Zixi of Ix,* are round like a ball and attack enemies by rolling at them. In *John Dough and the Cherub,* one of the main characters is Para Bruin, a large rubber bear that likes to roll into a rubber ball and bounce around.

Borges, writing about animals in the form of spheres, tells us that Plato, in the *Laws,* conjectures that the earth, planets and stars are alive. The notion that the earth is a living, breathing organism was later defended by such mystics as Giordano Bruno, Kepler, the German psychologist Gustav Theodor Fechner and Rudolf Steiner (who broke away from theosophy to found his rival cult of anthroposophy). The same notion is basic to the plot of one of Conan Doyle's stories about Professor George Edward Challenger of *Lost World* fame. When Professor Challenger drills a deep hole through the earth's epidermis, in a story called "When the Earth Screamed," the planet howls with pain.

Rotating wheels and propellers are common mechanisms for transporting man-made vehicles across ground, and through the sea and the air, but until recently it was assumed that evolution had been unable to exploit rotational devices for propulsion. Biologists were amazed to discover that the flagella of bacteria actually spin like propellers (see the article by Howard C. Berg).

The imaginary wing of our zoo would display two of Baum's creatures that use the wheel for propulsion. In *Ozma of Oz* Dorothy has an unpleasant encounter with the Wheelers, a race of fierce, four-legged humanoids that have wheels instead of feet (see Figure 20). In *The Scarecrow of Oz* we read about the Ork, a huge bird with a propeller at the tip of its tail (see Figure 21). The propeller can spin both ways, enabling the bird to fly backward as well as forward.

I know of only two imaginary beasts that bend themselves into wheels and roll across the ground. From time to time, in most parts of the world, people have claimed to have seen

FIGURE 20 A Wheeler

FIGURE 21 The Ork

"hoop snakes" that bite their tails to form a hoop and then go rolling across the terrain. Some snakes, such as the American milk snake, travel by gathering their body into large vertical loops and pushing forward so rapidly that they create an optical illusion of a rolling ring. These animals may be the origin of hoop-snake fables.

The Dutch artist M. C. Escher made several pictures featuring his curl-up, the beast shown in Figure 22. This unlikely animal moves slowly on six humanlike feet, but when it wants to go faster it curls up and rolls like a wheel.

Most animals, particularly the earthworm, may be thought of as being basically toroidal—a shape topologically equivalent to a doughnut. There must be many science-fiction animals shaped like toruses, but I can recall only the undulating silver ringfish, floating on the canals of Ray Bradbury's *Martian Chronicles,* that closes like an eye's iris around food particles.

Topologists know that any torus can be turned inside out through a hole in its surface. There is no parallel in earth

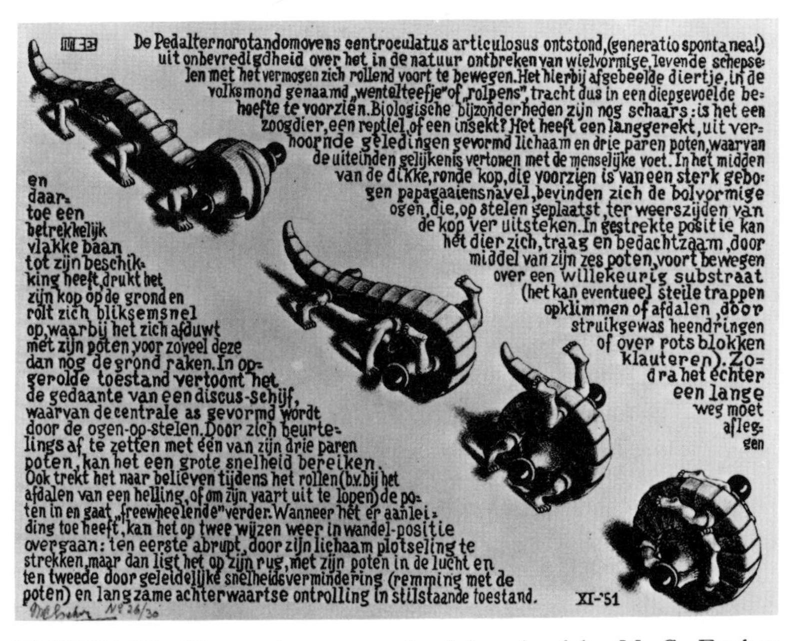

FIGURE 22 The curl-up, an animal imagined by M. C. Escher, can roll like a wheel when it wants to. (Collection Hoags Gemeente-museum–the Hague.)

zoology, but there is a spherical organism called volvox that actually does turn inside out through a hole. It is a strange freshwater-pond colony of hundreds of flagellated cells bound together in a spherical jellylike mass that rotates as it moves through the water. Volvox is one of those twilight things that can be called a green plant (because it obtains food by photosynthesis) or an animal (because it moves freely about). One is equally hard put to decide whether it is a colony or a single organism.

Young volvox colonies grow inside the mother sphere, but the cells have their flagella ends pointing inward. At the spot where each infant sphere is attached to the inside of the mother, there is a small hole in the infant sphere. When the infant reaches a certain size, it breaks away from the mother and turns inside out through the hole! Flagella quickly sprout at the ends of the cells that now point outward, and the newborn colony goes spinning about inside the mother. The mother eventually dies by splitting open and allowing her offspring to escape, one of the earliest examples on the evolutionary tree of nonaccidental death (see the article by John Tyler Bonner).

We could have considered volvox earlier, but I kept it for now to introduce the ta-ta, a mythical but much higher form of animal capable of turning inside out. It was invented by Sidney H. Sime, the British artist who so wondrously illustrated Lord Dunsany's fantasies. Sime drew and described the ta-ta in his only book, *Bogey Beasts,* a rare collection of original verses set to music:

The Ta-Ta

There is a cosy Kitchen
Inside his roomy head
Also a tiny bedroom
In which he goes to bed.

So when his walk is ended
And he no more would roam
Inside out he turns himself
To find himself at Home.

He cleared away his brain stuff
Got pots and pans galore!
Sofas, chairs, and tables,
And carpets for the floor.

He found his brains were useless,
As many others would

If they but tried to use them
A great unlikelihood.

He pays no rent, no taxes
No use has he for pelf
Infested not with servants
He plays with work himself.

And when his chores are ended
And he would walk about,
Outside in he turns himself
To get himself turned out.

ANSWERS

Figure 23 shows the answer to the first problem: an eight-sided deltahedron (all faces equilateral triangles) that is not a regular octahedron. The regular octahedron has four edges meeting at each corner. On this solid two corners are meeting spots for three edges, two for four edges and two for five edges.

The second problem was to use Euler's formula, $F+C-E=2$, to show that no sphere can be covered with a "regular map" of hexagons, each vertex the meeting point of three edges. Assume such a map exists. Each hexagon has six edges and six corners. Therefore if the hexagons did not share corners and edges, there would be six times as many edges as faces. Each corner is shared, however, by three faces;

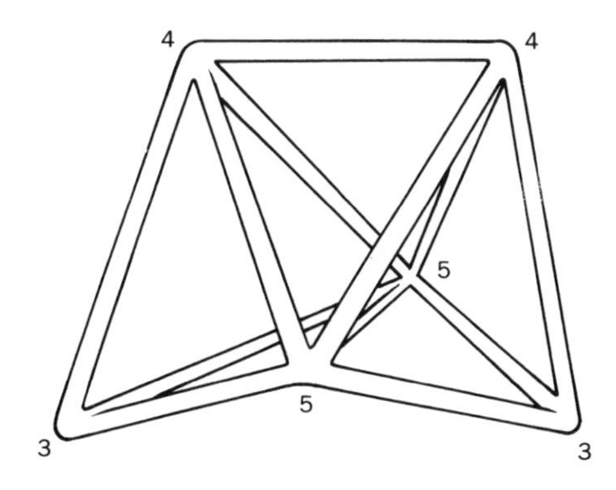

FIGURE 23 The "other" eight-sided deltahedron

therefore the number of corners in such a map must be $6F/3$. Similarly, each edge is shared by two faces; therefore the number of corners in such a map must be $6F/2$. Substituting these values in Euler's formula gives the equation $F+6F/3-6F/2=2$, which simplifies to $F+2F-3F=2$, or $0=2$. This contradiction proves the original assumption to be false.

What happens when the above argument is applied to the regular maps formed by the edges of the five Platonic solids? In each case we get a formula that gives F a unique value: 4, 8 and 20 for the tetrahedron, octahedron and icosahedron respectively, 6 for the cube and 12 for the dodecahedron. Since a regular polyhedron cannot have faces with more than six edges, we have proved that no more than five regular solids can exist.

Euler's formula also underlies an elementary proof that there are exactly eight convex deltahedra. See the paper by Beck, Bleicher, and Crowe cited in the bibliography.

ADDENDUM

I was mistaken in saying that no animal propels itself across the ground by rolling like a disk or sphere. Brier Lielst, Philip Schultz, and a geologist with the appropriate name of Paul Pushcar, were among many who informed me of a *National Geographic* television special on March 6, 1978, about the Namib desert of Africa. It showed a small spider that lives in burrows in the sides of sand dunes. When attacked by a wasp, it extends its legs like the spokes of a wheel and escapes by rolling down the dune.

Peter G. Trei of Belgium sent me a copy of a note in *Journal of Mammalogy* (February, 1975) by Richard R. Tenaza, an American zoologist. Tenaza describes how on Siberut, an island west of Sumatra, he witnessed the technique by which pangolins, a species of scaly anteater, elude capture: they curl themselves into a tight ball and roll rapidly down a steep slope (see Figure 24). In fact, the name "pangolin" is from a Malay word meaning "to roll."

Roy L. Caldwell, a zoologist at the University of California, Berkeley, wrote to me about an inch-long crustacean called *Nannosquilla decemspinosa,* found in the sands off the coast of

FIGURE 24 An uncurled pangolin

Panama. "When exposed on dry land, the animal's short legs are not sufficiently strong to drag its long, slender body, so it flips over on its back, brings its tail up over on its head, and takes off rolling much like a tank-track. While it does not actually close the perimeter by grasping the tail in its mouthparts, they are usually kept in close proximity. The path taken is usually a straight line, the animal can actually climb a five-degree slope, and it can make a speed of about six cm/sec."

Thomas H. Hay told me about a species of wood lice (also called slaters, sow-bugs, and pill-bugs) that, when alarmed, curl into a ball and roll away. Hay said his children call them roly-polys. While working underneath his car, with a trouble light beside him, he often finds the roly-polys "advancing in inexorable attack. I fantasize that they are Martian armored vehicles, released from a tiny spacecraft. Fortunately, they move so slowly that my work is finished before they constitute a threat."

Robert G. Rogers, in a letter that appeared in *Discover* (October, 1983) had this interesting comment on an earlier *Discover* article, "Why Animals Run on Legs, Not on Wheels" (see the bibliography):

> The concept of animals developing wheels for locomotion is not so far-fetched. A wheel with a diameter of one foot has a circumference a little over three feet. If it were mounted on a bone-bearing joint, with flexible veins and arteries, and a continuous series of circumferential pads (as on a dog's paw), the wheel could be wound back one turn by its internal muscles,

then placed on the ground and rotated forward two full turns, traveling about six and a quarter feet. While one wheel (or pair of wheels in a four-legged animal) is driving the creature along the ground, the other would be lifted up and rotated back in preparation for its next turn at propulsion. At a speed of ten m.p.h., the creature would be traveling about 15 feet per second—not an impossible pace.

Matthew Hodgart, writing from England, reminded me that the human animal is capable of moving by repeated somersaults, cartwheels, forward and back flips, and that two persons can grab each other's feet and roll like a hoop. Hodgart quoted these lines from Andrew Marvell's poem "To a Coy Mistress":

Let us roll all our strength, and all
Our sweetness, up into one ball:
And tear our pleasures with rough strife,
Through the iron gates of Life.

"I don't quite know what's going on here," Hodgart adds.

Chandler Davis supplemented my list of imaginary creatures that roll by calling attention to such an animal in George MacDonald's fantasy *The Princess and Curdie.* Ian F. Rennie thought I should have mentioned the Wumpetty-Dumps, found in *The Log of the Ark,* by Kenneth Walker and Geoffrey Boumphrey.

Is the narwhal the only animal with a name starting with N? Garth Slade cited the numbat, a small marsupial that lives in Western Australia. An article in *Word Ways* (May, 1973) gave two other examples: the nutria, a web-footed South American aquatic rodent (now also flourishing on the Gulf Coast and the coasts of the Pacific northwest), and the nilgai, an antelope in India that is commonly called a "blue bull" because of its bluish-gray color. It is curious that the best one can do with common names are the colloquial nag and nanny-goat.

Arthur C. Statter sent his reasons for thinking that the drawing by Haeckel which is reproduced in Figure 16 (I picked it up from D'Arcy Thompson's *On Growth and Form*), was one of many drawings that Haeckel deliberately faked. The forms shown, Statter says, simply do not exist. I have not tried to investigate this, and would welcome opinions from radiolaria experts.

Rufus P. Isaacs, commenting on the impossibility of tessellating a sphere with hexagons, sent a proof of a surprising theorem he discovered many years earlier. If a sphere is tessellated with hexagons and pentagons, there must be exactly 12 pentagons, no more and no fewer.

A soccerball is tessellated with twenty hexagonal "faces" and twelve that are pentagons. In 1989 chemists succeeded in creating the world's tiniest soccerball—a carbon molecule with 60 atoms at the vertices of a spherical structure exactly like that of a soccerball (see Figure 25). It is called a buckeyball, or more technically, a buckminsterfullerene, after its resemblance to Buckminster Fuller's famous geodesic domes. It belongs to a class of highly symmetrical molecules called fullerenes.

The buckeyball is known to geometers as a truncated icosahedron because it can be constructed by slicing off the 12 corners of a regular icosahedron. No one yet knows what properties this third form of carbon (the other two are graphite and diamond) might have. Because of the molecule's near spherical shape, it might provide a marvelous lubricant. (See "Buckeyball: The Magic Molecule," by Edward Edelson, in *Popular Science,* August, 1991, page 52ff.)

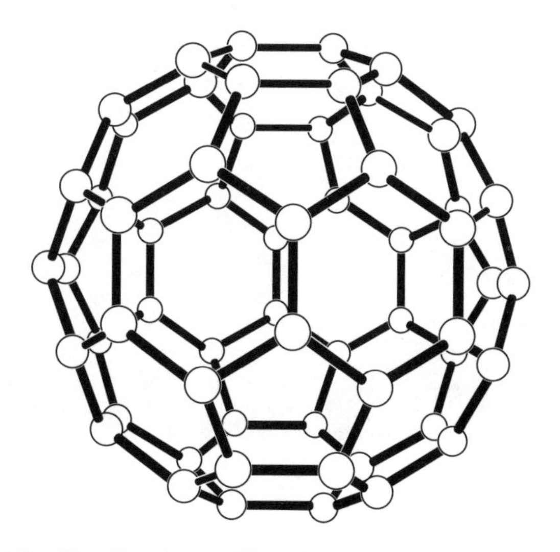

FIGURE 25 The "buckeyball" molecule—the world's smallest soccerball.

FIGURE 26 Two views of the "deceptahedron," an 18-sided deltahedron that is almost convex

It also can be shown that if a sphere is tessellated with hexagons and triangles, there must be an even number of hexagons and exactly four triangles. These results suggest the following general question: What are the integral values of *k* such that the sphere can be tessellated with hexagons and exactly *k* polygons of side *n?* As far as I know, this question has not been completely answered, though many special cases have been proved.

Emerson Frost sent photographs of his paper models of the eight convex deltahedra, as well as many of the nonconvex forms. Figure 26 shows his model of the 18-sided deltahedron that is so close to being convex that William McGovern (see the bibliography) has dubbed it the "deceptahedron." It is a pity that teachers are not as familiar with the eight convex deltahedra as they are with the five Platonic solids because constructing models and proving the set unique by way of Euler's formula are splendid classroom challenges.

BIBLIOGRAPHY

"Volvox: A Colony of Cells." John Tyler Bonner, in *Scientific American,* May, 1950, pages 52–55.

"The Horn of the Unicorn." John Tyler Bonner, in *Scientific American,* March, 1951, pages 42–46.

"The Structure of Viruses." R. W. Horne, in *Scientific American,* January, 1963, pages 48–52.

"Knots in Leucothrix Mucor." Thomas D. Brock, in *Science,* 144, 1964, pages 870–871.

"The Hagfish." David Jensen, in *Scientific American,* February, 1966, pages 82–90.

The Book of Imaginary Beings. Jorge Luis Borges. Dutton, 1969.

"How Snakes Move." Carl Gans, in *Scientific American,* June, 1970, pages 82–96.

Animal Asymmetry. A. C. Neville. Edwin Arnold, London, 1976.

The New Ambidextrous Universe. Martin Gardner. W. H. Freeman, 1989.

On The Bee Honeycomb

"The Mathematics of the Honeycomb." David F. Siemens, Jr., in *Mathematics Teacher,* April, 1965, pages 334–337.

"What the Bees Know and What They Don't Know." L. Fejes Tóth, in the *Bulletin of the American Mathematical Society,* 70, 1964, pages 468–486.

On Animal Wheels

"How Bacteria Swim." Harold C. Berg, in *Scientific American,* August, 1975, pages 36–44.

"Why Animals Run On Legs, Not On Wheels." Jared Diamond, in *Discover,* September, 1983, pages 64–67.

"Bacteria's Motors Work in Forward, Reverse, and Twiddle." Leo James, in *Smithsonian,* September, 1983, pages 127–134.

On Deltahedra

"Deltahedra." H. M. Cundy, in *Mathematical Gazette,* 36, 1952, pages 263–266.

"The Structure of Liquids." John Desmond Bernal, in *Scientific American,* August 1960, pages 124–130. The author maintains that molecules in liquids form deltahedral structures.

"Deltahedra." Anatole Beck and Donald Crowe, in *Excursions into Mathematics,* Anatole Beck, Michael Bleiler, and Donald Crowe, Worth, 1969, pages 21–26.

"A Recursive Approach to the Construction of the Deltahedra." William E. McGowan, in *Mathematics Teacher,* 71, 1978, pages 204–210.

"An Infinite Class of Deltahedra." Charles W. Trigg, in *Mathematics Magazine,* 51, 1978, pages 55–58.

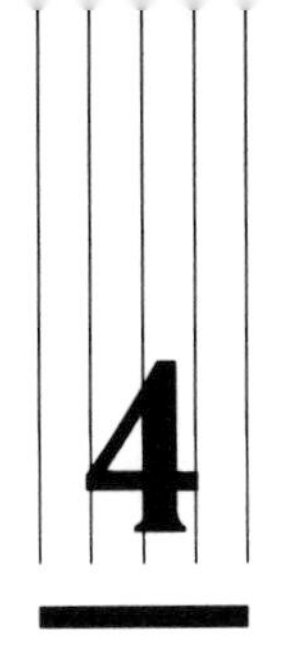

Charles Sanders Peirce

> I could make the whole matter clear to you as the noonday sun, if it were not that you are wedded to the theory that you can't understand mathematics!
>
> —From a letter of CHARLES SANDERS PEIRCE to WILLIAM JAMES

Most of the famous philosophers of the past had little talent for mathematics, but there are some notable exceptions. Descartes, Leibniz, Pascal, Whitehead, Russell—they are as eminent in the history of mathematics as in the history of philosophy. To this small, select band belongs Charles Sanders Peirce (1839–1914), scientist, mathematician, logician and the founder of pragmatism. In the opinion of many he was America's greatest philosopher.

Peirce was trained in mathematics by his father, Benjamin Peirce, the leading U.S. mathematician of his day, but of the two, Charles was by far the more original. His contributions to logic, the foundations of mathematics and scientific method, decision theory and probability theory were enormous. It is remarkable how many later developments he anticipated. At a time when the infinitesimals of early calculus were in disrepute, Peirce insisted on their usefulness, a view only recently vindicated by the invention of nonstandard analysis. At a time when determinism dominated physics Peirce's doctrine of "tychism" maintained that pure chance—events undetermined by prior causes—are basic to the universe. This is now essential to quantum mechanics. Even Peirce's notion that natural laws are "habits" acquired by a growing universe is no longer as eccentric as it once seemed. There are respectable models of oscillating universes in which random events create a different set of constants at each bounce. As each cosmos explodes it develops laws, some of which change as the universe cools and ages.

Peirce's influence on William James, a longtime friend whom he adored, was much greater than the other way around. The basic idea of pragmatism, including the word itself, was introduced by Peirce in a popular magazine article. James picked up the word and enlarged on Peirce's suggestions in a series of brilliant lectures that became the book *Pragmatism*. Peirce was so annoyed by what he considered James's reckless exaggerations that he changed his word to "pragmaticism," a term so ugly, he declared, that no one would kidnap it.

Like all creative mathematicians Peirce enjoyed mathematics hugely as a form of intellectual play. As a child he had had an intense interest in chess problems, puzzles, mathematical card tricks and secret codes. This sense of amusement runs through all his mature writings. He even coined the word "musement" for a mental state of free, unrestrained speculation, not quite as dreamy as reverie, in which the mind engages in "pure play" with ideas. Such a state of mind, he maintained, is the first stage in inventing a good scientific hypothesis. One meditates on all the relevant data, then pushes them around in one's head to form new combinations (like pushing chess pieces to solve a chess problem) until comes that mysterious flash of insight.

In a little-known paper titled "A Neglected Argument for the Reality of God" Peirce argued that "musement" is not only a road to theism but also the only road. It is a leap comparable to the scientist's conjecture, although it is one of the heart rather than of the head. It is not testable, but for those who make it, Peirce wrote, it can be as certain as the belief in one's own existence or the existence of others. It was on such matters of "over belief" (James's term) that Peirce and James agreed.

Peirce's recreational approach to mathematics is most evident in his views on how mathematics should be taught to children. He was convinced that the methods then in use produced only dunces. The manuscripts of his three unpublished textbooks are filled with novel ways of using puzzles, games and toys for introducing mathematical concepts. For example, Zeno's paradoxes lead into discussions of the continuum and the limit. Projective geometry and the shadows of a rotating wheel illuminated by a lamp introduce infinity. Peirce recognized—this before 1900!—the great value of elementary topology (he called topology the "easiest, most elementary and most fundamental branch of geometry") in stimulating a child's mathematical imagination. Euler's formula for the skeletons of polyhedrons, knot theory, graph theory, the four-color-map conjecture (which Peirce tried vainly for decades to prove), the Möbius strip—these are only some of the topological topics Peirce used to arouse student interest. He delighted in asking teachers to let him instruct a group of youngsters who detested mathematics and seemed incapable of learning it. He records that in one case, after about 10 lessons, two of his "prize stupids" led the school.

To teach arithmetic Peirce recommended the constant use of counters such as beans, the early introduction of binary notation, the use of 101 cards numbered 0 through 100 and other devices now common in grade school instruction. In one textbook he wanted to insert a cardboard mechanical gadget for doing multiplication. "The objection to inserting this," he jotted in a notebook, "would be that the teachers would not understand the mathematical principle on which it depends, and might therefore be exposed to embarrassing questions."

The use of playing cards is also recommended. "If you will provide yourself, my dear Barbara, with a complete pack of cards with a joker, 53 in all, I will make a little lesson in mathematics go down like castor-oil in milk." Barbara is a character in one of Peirce's unpublished textbooks. She is so named because "Barbara" was the medieval mnemonic name for the syllogism "All *A* is *B*, all *B* is *C*, therefore all *A* is *C*."

In the introduction to another textbook Peirce devotes 15 pages to ticktacktoe! The game is used for showing how a theorem is first guessed, then tested by manipulating diagrams. "Such are the tools," he writes, "with which the mathematician works." With the huge success of textbooks such as Harold R. Jacobs' *Mathematics, A Human Endeavor,* some teachers have caught up with the proposals in Peirce's unsalable manuscripts.

Like so many other mathematical geniuses—Leibniz and Kepler come to mind—Peirce sometimes became over-enthusiastic, almost obsessively so, about some of his creations. This may have been partly the result of his working alone, without the give and take of the classroom or constant discussion with colleagues. Peirce did not get along with most people, and in later years his ill-temper and poverty made him a lonely recluse. James described him as a "poor cuss" to whom no university would give a professorship, a "queer being," a "hopeless crank" and a man whose lectures were "flashes of brilliant light relieved against Cimmerian darkness!" In a touching tribute that Peirce wrote after James died, Peirce said, "Who . . . could be of a nature so different from his than I? He so concrete, so living; I a mere table of contents, so abstract, a very snarl of twine."

One of Peirce's two major obsessions was his system of "existential graphs" for diagramming logic. He was on solid ground in seeing the pedagogical value of Venn diagrams for solving syllogisms and more general problems in Boolean algebra, but he wanted to extend such visual aids to every kind of logic, including modal logic. His system grew steadily more complex, relying always on topological properties of the plane. For 20 years he used his curious diagrams as thinking aids, and there is no question that he found them enormously useful. He called them his *chef d'oeuvre,* and believed that if they were "taught to boys and girls before grammar . . . it would aid them through all their lives." If logicians would embrace his method, he wrote, "there would soon be such an advance in logic that every science would feel the benefit of it." Unfortunately no one else found the graphs useful, although it may be too early to give a final verdict. An excellent monograph by Don D. Roberts, *The Existential Graphs of Charles S. Peirce,* was published in 1972.

Peirce's other great eccentricity—perhaps I tread on even more dangerous ground in calling it that—was his conviction that in every branch of philosophy the most efficient way to organize concepts is by way of three fundamental categories that he called firstness, secondness and thirdness. Like scientists, philosophers are compelled to classify ideas, and since philosophy is about everything, their schemes often include a list of what they consider the most fundamental categories. Aristotle's 10 categories had such an enormous influence on Western philosophy that it was not until Kant proposed a different set that Aristotle's scheme met serious competition. Kant had 12 categories (in four triads) that he considered essential for describing how human consciousness imposes patterns on the vast, ultimately unknowable sea of being. Since Kant there have been so many different schemes that "category" has become a fuzzy and unfashionable word.

Peirce was firmly persuaded that the most useful of all philosophic tools was the ordering of things into monads, dyads and triads. Firstness considers a thing all by itself, for example redness. Not a red object, just the pure possibility of red: "Redness before anything in the universe was yet red." Secondness considers one thing in relation to another, for example a red apple. It is redness linked to an apple, a "brute fact" of the actual world. Thirdness concerns two things "mediated" by a third, for example an apple falling from a tree. The tree and the apple are linked by the relation "falling from." Our mental concept of a red apple is another thirdness because it involves apple, red and mind. The universe "out there," changing in time, and the inner world of consciousness are equally "real" realms of thirdness.

Peirce applied firstness, secondness and thirdness to every branch of philosophy. There is no need, he argued, to go on to fourthness, fifthness and so on, because in almost every case these higher relations can be reduced to combinations of firstness, secondness and thirdness. On the other hand, genuine thirdness can no more be reduced to secondness than can genuine secondness to firstness. Peirce modeled this notion with a clever bit of graph-theory sleight-of-hand. Let a point represent firstness and the end points of a line segment represent secondness. Thirdness is symbolized by the ends of three line segments meeting at a common point like the map of a forked road. Why not go on to four, five, six and so on by letting more lines join at a point? Because we can always reduce such higher "stars" to thirdness by substituting triadic graphs for the central point as shown in Figure 27. There is no way this can be done, however, to reduce a triadic graph to one with two end points.

Peirce regarded his three categories as his greatest contribution to philosophy. He denied the charge that he was in-

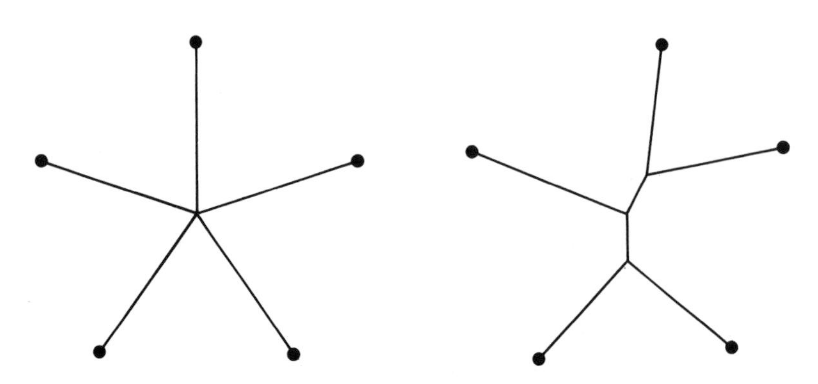

FIGURE 27 How Peirce reduced fifthness to thirdness

fatuated by the number three. He admitted having a Hegelian "leaning for 3," but he insisted that this was because thirdness had so many applications.

"The most fundamental fact about the number three," Peirce wrote, "is its generative potency. . . . So prolific is the triad in forms that one may easily conceive that all the variety and multiplicity of the universe springs from it." He called his graph for the triad "an emblem of fertility in comparison with which the holy phallus of religion's youth is a poor stick indeed." I do not know whether Peirce was familiar with the following statement by Laou-tsze, five hundred years before Christ: "*Tao* hath produced one, one hath produced two, two hath produced three, and three have produced all things."

This is not the place to discuss the fertility of Peirce's categories, and so I content myself with saying that in Peirce's day the only eminent philosopher who shared his enthusiasm for one, two and three was Josiah Royce. James complained that he never could understand Peirce's categories. Among today's philosophers of note I know of only two enthusiasts, Eugene Freeman and Charles Hartshorne. "I believe," Hartshorne has written, "that all things, from atoms to God, are really instances of First, Second, Third, and that no other equally simple doctrine has the power and precision of this one."

Let us turn to something less controversial: a card trick. In the April, 1908, issue of *The Monist* Peirce had an article on "Some Amazing Mazes" that opened with an apt description from Milton's *Paradise Lost* (Book V, 623–624) of a "mystical dance" of angels:

> *. . . Mazes intricate,*
> *Eccentric, intervolv'd, yet regular*
> *Then most, when most irregular*
> *they seem.*

"About 1860," Peirce begins, "I cooked up a *mélange* of effects of most of the elementary principles of cyclic arithmetic; and ever since, at the end of some evening's card-play, I have occasionally exhibited it in the form of a 'trick' . . . with the uniform result of interesting and surprising all the company, albeit their mathematical powers have ranged from a bare sufficiency for an altruistic tolerance of cards up to those of some of the mightiest mathematicians of the age, who assuredly with a little reflection could have unraveled the marvel."

By cyclic arithmetic Peirce meant what is today called congruence arithmetic. Some teachers call it "clock arithmetic" because it is so nicely modeled by a clock. For example, 2 is equal to 14 by modulo 12. This means that if you divide 2 and 14 by 12 (the modulus) the remainder in each case is 2. In clock terms, at 14 hours past noon the hands of the clock are in the same position as they are two hours from noon.

The first of Peirce's card tricks, reprinted in Volume 4 of his *Collected Papers* as "The First Curiosity," is surely the most complicated and fantastic card trick ever invented. I cannot recommend it for entertaining friends unless they have a passion for number theory, but for a teacher who wants to "motivate" student interest in congruence arithmetic it is superb. There is no way to prove that the trick always works without learning a great deal about "cyclic arithmetic," including a famous theorem of Fermat's about prime numbers.

Before reading further, the reader is urged to get a deck of cards and carefully follow the procedure. Remove all the hearts and arrange them in serial order from ace to king, the ace on top of a face-down packet. Do the same with spades, except for the king, which is not used. Thus the spade packet consists of 12 face-down cards from ace on the top to queen on the bottom. Put the red packet face-down on the table. Hold the black packet face-down in one hand.

Deal the black cards face-up onto two piles. (Whenever cards are dealt into piles they are held face-down and dealt face-up, from left to right, starting on the left.) The last card (the queen) is discarded by placing it face-up to one side to form a discard pile. Substitute for it the top card (the ace) of the red packet, putting it face-up on the second pile in place of the discarded queen. Assemble the two piles by picking up the pile farthest to the left and dropping it face-up on the second pile. Turn the black packet (now containing one red card) face-down and repeat exactly the same procedure. This time the red deuce replaces the last black card (the jack). The jack goes face-up on the previously discarded black queen. The procedure is continued until it has been performed 12 times in all. You may be surprised to discover that you now hold an all-red packet, and that the discard pile contains all the black cards. Pick up the remaining king of hearts and add it to the bottom of the face-down red pile.

To make sure you have done all this properly, check the red packet. Held face-down, and reading from the top, the order of cards should be: 7, 8, *J*, 9, 4, *Q*, 6, 10, 3, 5, 2, *A*, *K*. The black packet should be: *Q*, *J*, 9, 5, 10, 7, *A*, 2, 4, 8, 3, 6.

The two packets are correlated in a curious manner. The value of the card at the nth position from the top of either packet gives the position from the top of the other packet of a card with the value n. For example, where is the jack of spades? Counting the jack as 11, look at the 11th card in the red packet. It is a 2. Check the second card in the black packet. It is the jack of spades. Where is the five of hearts? The fifth card in the black packet tells you. It is a 10. The 10th card in the red packet is the five of hearts.

Before you reveal this remarkable correlation to your audience, however, the red packet is apparently randomized by the following procedure. First allow the packet to be cut as often as anyone wants. Hold it face-down and ask someone to name a number from 1 through 12. Call the number k. Deal the cards face-up into k piles, then assemble them by starting with any pile designated. The assembled packet can then be cut again and the procedure repeated as often as you like, either with the same k or a different one. One would suppose that cutting, dealing into k piles, assembling and cutting, and repeating this routine many times with any k requested, would hopelessly mix the red cards. Astonishingly, thanks to the theorems of congruence arithmetic, the correlation of the two packets is conserved!

The only difficult part of the mixing procedure is this. When the k piles are assembled, you must do it in a precise manner that depends on the value of k. Think of the row of piles as being circular, the last pile adjacent to the first, so that you can count "around" the row in a direction either clockwise or counterclockwise. Note the pile on which the last card was dealt. In your mind call the end pile on the right zero and count to the pile that got the last card. Count clockwise or counterclockwise, whichever is shorter.

Suppose you have dealt the red cards into five face-up piles. The middle pile will get the last card, as shown in Figure 28. It is second from the right, counting counterclockwise from the right end. This means that you must assemble the cards as follows. Pick up any pile and place it face-up on the second pile to the left, counting counterclockwise. Pick up the enlarged pile and place it on the second pile leftward, and continue until there is a single packet. The numbers under the piles in the illustration show the order of assembly if you start with the first pile on the left.

It is important to remember that in gathering the piles you count positions, not the actual piles. For some *k* the pile getting the last card is adjacent to the rightmost pile. This makes the assembly simple because the piles go on adjacent piles. But if there is a wider separation (as in the case of nine piles where the shortest distance is 4) it takes a bit of experience to assemble the piles rapidly. Figure 29 shows the order of picking up nine piles if you start with the second pile from the left. In this case you proceed clockwise because the shorter count from the right end to the pile that got the last card is a clockwise count. As you are practicing you can mark the positions with a row of pennies. After a while you can dispense with the markers.

The assembled packet can always be cut as many times

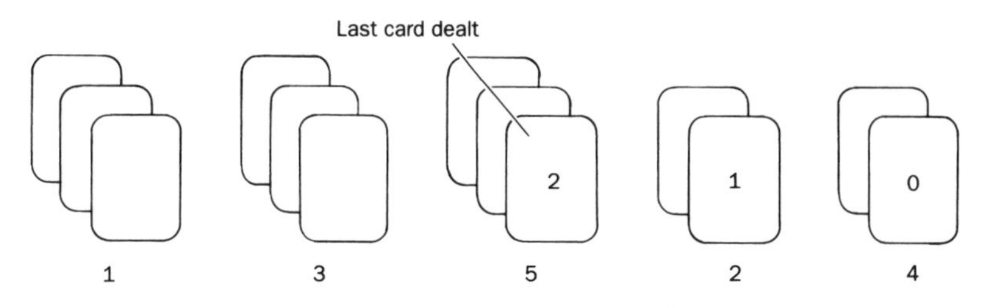

FIGURE 28 How to assemble five piles of cards in Peirce's card trick

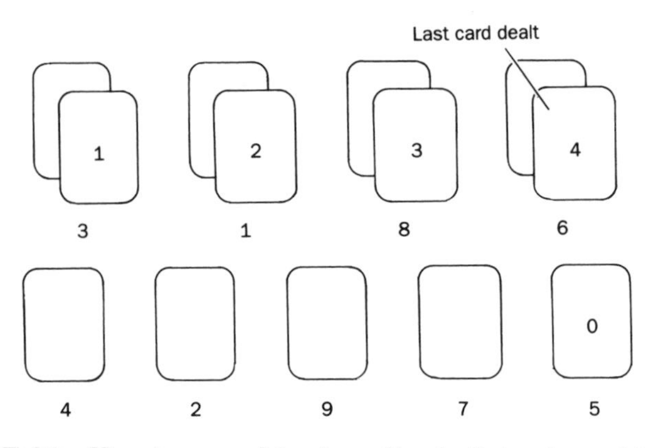

FIGURE 29 How to assemble nine piles in Peirce's card trick

as it is desired and the mixing repeated with a different number of piles. When everyone is satisfied that the red pile has been thoroughly "shuffled," it is necessary to give both the red and black piles a single cut. Cut the red pile to bring the king to the bottom when the pile is face-down. Peirce suggests remarking that since there is no king of spades, you will cut the red pile to bring the king to the bottom "and so render any searching for that card needless." As you do this, note the packet's top card. Suppose it is a 4. The black pile must now be cut so that its ace is fourth from the top. The two piles will then be correlated as before!

To dramatize the correlation Peirce suggests dealing the black cards in a face-down row. Ask anyone to name a red card. Suppose he says the 7 of hearts. Tap the black cards one at a time, counting from 1 to 7, and turn the seventh card face-up. Note its value *n*. Count to the *n*th card in the red packet. It will be the 7 of hearts.

"The company never fail to desire to see the thing done again," writes Peirce. After revealing a number of cards to be at the indicated positions, you can repeat the trick by mixing the red cards again with several deals, then adjust the two packets by the required cuts. "If you wish for an explanation of it," Peirce concludes, "the wish shows that you are not thoroughly grounded in cycle arithmetic." He refers his readers to a book by Richard Dedekind, but adds that on another occasion he may write a little essay on the topic.

This "half promise" he "half redeems," as he puts it, in an article in the July, 1908, issue of *The Monist,* also reprinted in the *Collected Papers*. Peirce's explanation of the trick runs to 58 pages! The essay is pure Peirce, complete with generalizations and formulas, horrendous existential graphs that look like abstract art and delightful digressions on such things as the value of keeping notes on file cards, logic machines, the meaning of continuity, how mind influences matter and the nature of free will and time.

In 1958 Alex Elmsley, a London magician, pointed out in *Ibidem* (a Canadian magic periodical) that in the first phase of Peirce's trick it is not essential that the last black card each time be the one that is replaced with a red card. The card to be replaced can be at any position in the packet. Thus you may allow someone to choose any number *n,* from 1 through 12, then on each deal you replace the *n*th card with a red one.

Peirce's writings are now gathered into 13 volumes, six edited by Charles Hartshorne and Paul Weiss, two by Arthur W. Burks and five by Carolyn Eisele. It is a triadic scandal (1) that the bulk of Peirce's mathematical papers were not published until 1976, when Mrs. Eisele skillfully put them together for *The New Elements of Mathematics,* (2) that these books have received almost no advertising or reviews and (3) that the set costs more than $400.

ADDENDUM

Many great philosophers have had a compulsion to carve experience into what they consider its most fundamental aspects, basic genera that have nothing in common with one another. Aristotle introduced the term "category" to refer to his ten basic divisions. For Kant there were four basic categories, each composed of triads, making twelve categories in all. Hegel found more than 200, but liked to divide all historical change into thesis, antithesis, and synthesis. The lower integers from 1 through 5 have always been the most popular for the simple reason that they are the simplest.

Consider oneness. Pre-Socratics liked to think of everything as made of one substance: earth, air, fire, or water. "All is one" is a favorite notion of pantheists and Eastern religions. For materialists, everything is matter. For idealists, everything is mind.

Twoness is another favorite. Being and nothing, mind and matter, yin and yang, body and soul, God and the universe, and so on.

Threeness, too, has bedazzled thinkers. A delightful summary of threeness can be found under the heading of "Trinity" in W. V. Quine's recent book *Quiddities*. "A predeliction for threes," he begins, "has invested song and story. We have the Three Fates, the Three Graces, the Three Magi, the Three Musketeers (actually four), the Three Bears, the Three Little Maids from School."

Quine goes on to cite Kant's various triads, Hegel's three movements of history, Peirce's triads, and Charles Morris's division of semiotics (theory of signs) into syntax, semantics, and pragmatics, a division taken over by Rudolph Carnap. Mathematical philosophy, Quine continues, is either formalism, logicism, or intuitionism. We have such phrases as liberty, equality, and fraternity; life, liberty, and the pursuit of happiness; faith, hope, and charity; and endless other triplets. In religion we have the Christian trinity, though the "tenuous" quality of the Holy Ghost suggests "trinity for trinity's sake." The Hindu trinity of Brahma, Siva, and Vishnu, I should add, doesn't count because above them is the unknowable Brahman.

"There is something stable and comforting about threeness," Quine writes, "that may explain its popularity." A stool can't have two legs, and four legs are wobbly on an uneven floor, but a three-legged stool is always stable. This is because three points always lie on a plane, whereas two points lie on many planes, and four need not lie on a plane. Moreover, Quine adds, models of triangles are rigid whereas models of higher polygons flex at their corners.

Peirce was wild about threes. All sorts of triads lace his speculations. Exactly what he meant by firstness, secondness, and thirdness is seldom clear because he kept changing his mind about them. In *The Century Dictionary,* firstness, secondness and thirdness are among the hundreds of mathematical terms defined by Peirce. Firstness is anything that can't be reduced to more basic parts. Secondness is a dyadic relation that can't be reduced to oneness. Thirdness is a triadic relation that can't be reduced to twoness.

The best introduction to all this is Book III of the first volume of Peirce's *Collected Papers*. It is a rough draft for a book to be called *A Guess at the Riddle* in which Peirce intended to apply 1, 2, 3 to everything, including theology. "This book, if ever written, as it soon will be if I am in a situation to do it, will be one of the births of time."

Peirce recognized "that higher numbers may present interesting special configurations from which notions may be derived of more or less general applicability." Different numbers have their champions: "Two was extolled by Peter Ramus, Four by Pythagoras, Five by Sir Thomas Browne, and so on. For my part, I am a determined foe of no innocent number; I respect and esteem them all in their several ways; but I am forced to confess to a leaning to the number Three." Again: "Other numbers have been objects of predeliction to this philosopher and that, but three has been prominent at all times and with all schools."

Peirce struggled mightily to show that relations involving four or more things can always be reduced to triads. No number of straight roads put end to end can have more than two ends, he wrote, but a road with any number of forks can (as previously explained) be built out of triads. Another example concerns A selling C to B for a price of D. This may look like a quadrad, but for Peirce it was a compound of two triads that involve a common event E, namely the sale. Thus E relates seller A to buyer B, and E also relates the object C to its price D.

"Every higher number," Peirce wrote, "can be formed by mere complications of threes." Again: "Any number, however large, can be built out of triads, and consequently no idea can be involved in such a number, radically different from the idea of three." What on earth does this mean? It is trivially true that 7 can be expressed by

$$\frac{3+3+3+3+3+3+3}{3}$$

but this formula gives 7 when any positive integer n is substituted for each 3. In similar ways any integer can be represented by combinations of any other integer.

Peirce admitted that during early stages of his work he would have considered the book he wanted to write "too strong a resemblance to many a crack-brained book that I had laughed over," but with deeper study he came to appreciate the power of his triads. In the opinion of almost everybody else except a handful of dedicated Peircians, Peirce's youthful attitude toward his book was the correct one.

It is easy to think of relations involving more than three things that cannot, in any reasonable way, be reduced to thirdness. Consider the complete graph of four points—a graph modeled by the edges of a tetrahedron. Each point is joined to the other three. Such a graph may symbolize a variety of unreducable fourthness. Four stars can form a gravitational system. The movement of any one star is governed by the movements and masses of the other three. Put another way, this four-body system is radically altered if any star is removed. For another example, consider a family of father, mother, son, and daughter. Their behavior obviously depends on how the four interact. Everything changes if one member of the family dies. The relations of four nations can't be reduced to threeness, nor can the four lines of a quatrain of poetry, the four notes of a chord, the many notes of a melody, four figures in a painting, the four legs of an elephant, or the six legs of a spider.

The only reason for stopping at three is that relations beyond three get more complicated. Two is company, three a crowd. There is indeed a big jump from two to three, as Peirce perceived, but there are also big jumps from three to four, or four to five. In life, in the universe, and in ideas there are clusters of any number of things so bound together that any subset is mediated by all the other elements. The universe is just too rich to force into 1, 2, or 3.

The late Bruno Bettelheim, an unbending Freudian, wrote a curious book about fairy tales called *The Uses of Enchantment*. In analyzing "Goldilocks and the Three Bears" Bettleheim revealed the number that he said represents sex in the unconscious mind. Is this mysterious number 1, based on the fact that every person has just one sex organ? No. Is it two, based on the fact that it takes two to tango? No. The number is 3! Why? Because, Bettleheim assures us, "each sex has three visible sex characteristics: penis and the two testes in the male; vagina and the two breasts in the female." That is the hidden meaning of the three bears in the familiar story. When Goldilocks peeked at the three bowls of soup, the three chairs, and the three beds, the tale "evokes associations to the child's desire to find out the sexual secrets of adults."

This is neither science nor literary criticism, but crackpot psychoanalysis. Peirce's attempt to reduce everything to

1, 2, or 3 is almost as bad. Like his existential graphs, it is an eccentric phase of his thinking that is best forgotten.

BIBLIOGRAPHY

On Peirce's Three Categories

Principles of Philosophy: Collected Papers, Vol. 1, Book III. Charles Sanders Peirce. Edited by Charles Hartshorne and Paul Weiss, Harvard University Press, 1931.

The Categories of Charles Peirce. Eugene Freeman. Open Court, 1934.

The Phenomenology of Charles S. Peirce: From the Doctrine of Categories to Phaneroscopy. William L. Rosensohn. B. R. Gruner, 1974.

"A Revision of Peirce's Categories." Charles Hartshorne, in *The Monist,* 63, 1980, pages 278–289.

On Peirce's Card Trick

The Simplest Mathematics: Collected Papers, Vol. 4. Charles Sanders Peirce. Edited by Charles Hartshorne and Paul Weiss, Harvard University Press, 1933, pages 473–549.

"Number-Theoretic Analysis and Extension of 'The Most Complicated and Fantastic Card Trick Ever Invented'." Kurt Eiseman, in *American Mathematical Monthly,* 91, 1984, pages 284–289.

"A Self-Dual Card Trick Based on Congruences and *k*-Shuffles." Kurt Eiseman, in *American Mathematical Monthly,* 93, 1986, pages 201–205.

5

Twisted Prismatic Rings

The well-known Möbius band, formed by giving a paper strip a half twist and joining the ends, is a model of an abstract surface of zero thickness. No paper, however, is zero thick. The cross section of a Möbius band actually is a rectangle, very much longer than it is wide. The band itself may be regarded as a four-sided prism, twisted so that one end has rotated 180 degrees before the two ends are joined. Viewed in this way it is a solid ring with two distinct "faces." One face is the flat surface of the band, which circles the ring twice. The other is the band's narrow but also flat edge, which also circles the ring twice.

Over the years I have received dozens of letters from readers who independently noticed that models of Möbius bands actually are twisted prisms and who generalized such rings by considering prisms with cross sections that are regular polygons with any number of sides. Although such structures have many strange properties, surprisingly little is known about them.

There is no agreed-on name for these structures, so let us call them prismatic rings. Let n be the number of sides of the polygon cross section, and let k be the number of $1/n$ turns the prism is given before its ends are joined. If the prism is not twisted, then $k=0$. If it is twisted (in either direction) so that each side joins an adjacent side, then $k=1$. If each side joins the next side but one, then $k=2$, and so on.

The easiest prismatic ring to visualize is the ring with a square cross section. If $k=0$ (no twists), the ring obviously has four sides and four edges. If $k=1$, we get the beautiful solid in Figure 30. (It is a photograph of a wood carving by Roger I. Canfield, who sent it to me after reading my December 1968 column, in *Scientific American,* on Möbius bands.) Like the Möbius surface, it has only one "face" and only one edge.

FIGURE 30 A twisted prismatic ring with one face and one edge

A 1949 science-fiction story by Theodore Sturgeon (see the bibliography) involved a man who was given an interesting intelligence test. He was placed inside an $n=4$, $k=1$ prismatic ring where artificial gravity fields kept him attached to the "floor." The test was to determine if he was smart enough to deduce the structure of the corridor as he walked around it.

When $n=4$ and the ring is given two twists, it becomes topologically equivalent to the familiar Möbius band viewed as a solid: two-faced and two-edged. Three twists (like one twist the other way) produce one face and one edge, and four twists bring the structure back to four faces and four edges. On all prismatic rings the faces equal the edges in number, so henceforth we shall consider faces only. The sequence of numbers for the faces repeats periodically when the twists exceed n. Thus a square prism with five twists has the same number of faces as one with a single twist. Note also that all twisted rings are mirror-asymmetric and therefore have mirror-image counterparts.

Let us generalize to prismatic rings with cross sections that are regular polygons of n sides. Given k twists it is easy to predict the number of faces. It is the GCD (greatest common divisor) of n and k. From this fact several interesting properties follow. If the cross section has a prime number of sides, the number of faces is n only when k (the number of twists) is 0 or any multiple of n. Otherwise the ring has only one face. If n is not prime, the ring has one face only when n and k are prime to each other (have no common divisor). The table in Figure 31 (sent to me in 1964 by John Steefel) gives the number of faces on prismatic rings with cross sections that have two through 15 sides and that have been given zero through 15 twists. Note that the Möbius strip appears here as a degenerate prismatic ring with a two-sided cross section.

Now the fun begins. We all know that crazy things happen to twisted bands when they are cut down the middle. Equally crazy things happen when twisted prismatic rings are cut in various ways. Figure 32, top, shows eight ways a ring with a square cross section can be cut.

Consider the first way, a simple bisection down the middle. If $k=0$, the result obviously is two separate rings, each with four faces and no twists. If $k=1$, the cutting goes twice around the ring, and the result is the same as if two perpendicular cuts are made as shown in the second diagram. This opera-

n = Sides of cross section \ k = Number of twists	0	1	2	3	4	5	6	7	8	9	10	11	12	13	14	15
2	2	1	2	1	2	1	2	1	2	1	2	1	2	1	2	1
3	3	1	1	3	1	1	3	1	1	3	1	1	3	1	1	3
4	4	1	2	1	4	1	2	1	4	1	2	1	4	1	2	1
5	5	1	1	1	1	5	1	1	1	1	5	1	1	1	1	5
6	6	1	2	3	2	1	6	1	2	3	2	1	6	1	2	3
7	7	1	1	1	1	1	1	7	1	1	1	1	1	1	7	1
8	8	1	2	1	4	1	2	1	8	1	2	1	4	1	2	1
9	9	1	1	3	1	1	3	1	1	9	1	1	3	1	1	3
10	10	1	2	1	2	5	2	1	2	1	10	1	2	1	2	5
11	11	1	1	1	1	1	1	1	1	1	1	11	1	1	1	1
12	12	1	2	3	4	1	6	1	4	3	2	1	12	1	2	3
13	13	1	1	1	1	1	1	1	1	1	1	1	1	13	1	1
14	14	1	2	1	2	1	2	7	2	1	2	1	2	1	14	1
15	15	1	1	3	1	5	3	1	1	3	5	1	3	1	1	15

FIGURE 31 A table showing the number of faces when n and k are known

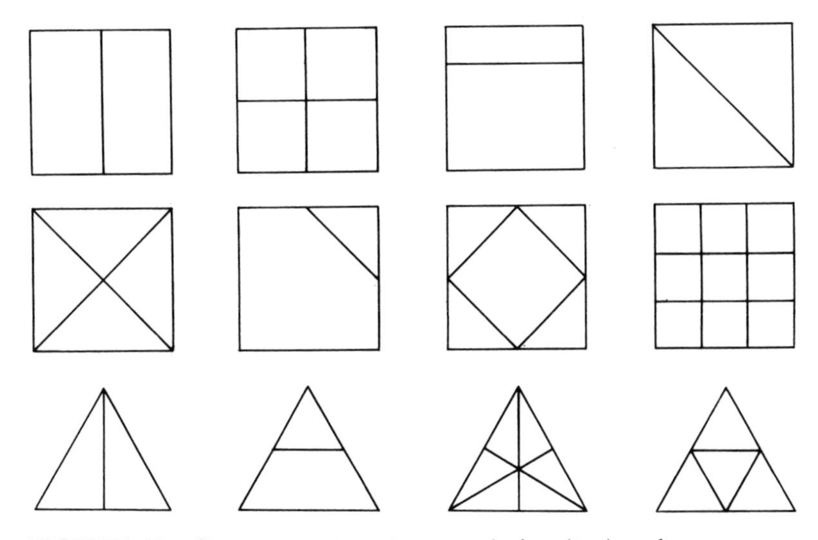

FIGURE 32 Some ways to cut an $n=4$ ring (top) and some ways to cut an $n=3$ ring

tion forms a single ring four times as long as the original, with four faces and 16 twists of 90 degrees each.

When $k=2$, the cutting forms a single ring with four faces and eight twists, but now the ring is only twice the size of the original. When $k=3$, the cutting goes around a second time to produce a knotted single ring four times as long as the original. It has four faces and 24 twists. Two interlocked rings result when $k=4$, each the same size as the original and each with four faces and four twists.

The cases of $k=2$ and $k=4$ can be modeled with paper strips, viewing them as solids. As we have seen, when $k=2$, we have the familiar Möbius band, and $k=4$ is a paper band with two half twists. Simply cut the strips down the middle, then examine the results, remembering that the edges are considered faces. To experiment with $k=1$ or $k=3$ an actual solid model is helpful. The simplest way to make such a model is with "salt ceramic": a mixture of one cup of table salt, half a cup of cornstarch and three-fourths of a cup of cold water. Put it in a double boiler, heat it and stir until it thickens and follows the spoon. Let it cool on wax paper. Knead out the lumps and shape the substance into strips about eight inches long, with four faces about half an inch wide. Form into the desired ring, smooth out the cracks with water and let dry. If you like, you can paint each face a different color before you start cutting.

As we have seen, when $k=1$ or $k=3$, the methods of cutting in the first two diagrams are the same. When $k=2$ or $k=4$, they are not the same. Here again we can model the double cut of the second diagram easily with paper strips. Simply put one strip on another and give the double strip either one half twist or two before joining the edges. Such strips, viewed as solids, are the same as rings with two or four quarter twists that have been bisected. Cutting these rings down the middle is then the same as adding a second cut perpendicular to the first one. In this way we see that if $k=2$, and the ring is bisected both ways, the result is two interlocked rings (one ring is twisted twice around the other), each four-faced, twice as long as the original and twisted eight times. If $k=4$, the result is four rings, all interlocked. Each is the same size as the original, with four faces and four twists.

Interested readers may wish to experiment with some higher values for k. If $k=6$, for example, and the ring is bisected, the result is a single ring tied in an overhand knot.

Also open to exploration are the other ways of cutting the $n=4$ ring, and the cutting of rings with cross sections that are other than square. Triangular-cross-section rings are easily made with salt ceramic, and Figure 32, bottom, shows four ways to cut them. Rings with pentagonal and still more complex cross sections are too difficult to model and cut, but they can be investigated, as Matthews explains, by using appropriate diagrams.

Are there formulas that, given n and k and the method of cutting, will predict the number of twists in the resulting ring or rings? Undoubtedly there are, but I know of nothing published along such lines.

So far we have considered only prismatic rings bent into circular shapes. We can, however, form them in such a way that they are toroidal polyhedrons. What we have called a ring's "face" will then consist of flat four-sided polygons that are joined along their edges in a circular chain. For example, Figure 33 shows a prismatic ring ($n=4$) of this type. The illustration is based on a photograph in a 1974 article in Spanish on such structures by Gonzalo Vélez Jahn of the architecture department of the Central University of Venezuela. The ring has a single twist; therefore it models a prismatic ring of square cross section that has one face (in our former meaning of "face") and one edge.

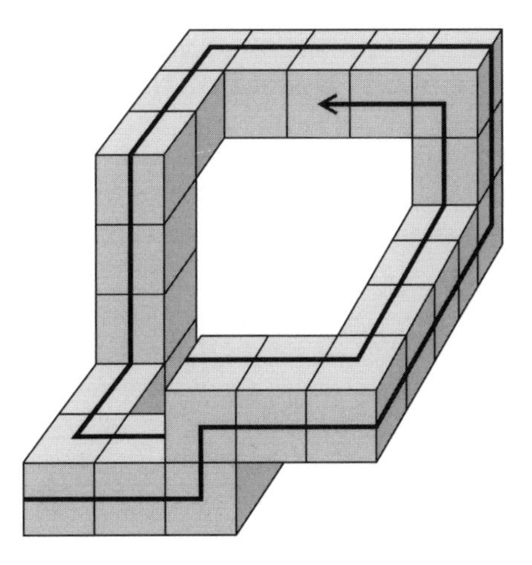

FIGURE 33 A prismatic one-sided ring made with 22 cubes

Such polyhedral prismatic rings suggest a variety of difficult problems that are only beginning to be explored. Scott Kim, for example, has proved a number of remarkable theorems about polyhedral rings. (Kim is an American mathematician best known for his book *Inversions,* W. H. Freeman, 1989.) They are closely related to a class of "impossible objects," such as the Penrose triangle shown at the left in Figure 34, and its rectangular version on the right.

It is conceivable that figures such as these are not really impossible but are drawings of twisted polyhedral prismatic ($n=4$) rings. D. A. Huffman, in his paper "Impossible Objects as Nonsense Sentences," was the first to devise algorithms for proving whether or not such figures are possible.

There are informal proofs for specific figures. Here, for instance, is Huffman's impossibility proof for the Penrose triangle. We first make some reasonable assumptions:

1. The straight lines in the drawing are straight lines on the actual model.

2. Regions that appear flat in the drawing actually are plane surfaces.

3. Surfaces A and B intersect at line 1, surfaces B and C intersect at line 2 and surfaces C and A intersect at line 3.

Three planes, no two of them parallel, will either intersect along three parallel lines, or intersect at a common point P. Therefore each of the three intersection lines must pass through P. Note, however, that lines 1, 2 and 3—the three lines of intersection—cannot meet at one point. Therefore the figure is not possible. It is amazing that this simple proof re-

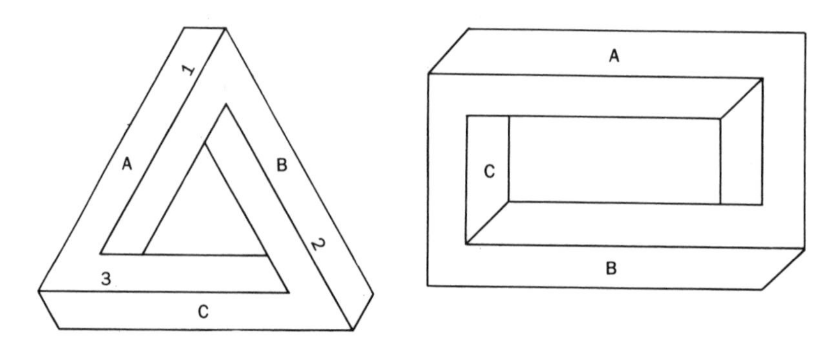

FIGURE 34 Two impossible polyhedral prismatic rings

quires no information about the hidden faces or edges of the picture. An impossibility proof for the rectangular "window" is even simpler because only sides *A* and *B* need be considered. Face *C* is irrelevant.

Kim called my attention to the solid shown in Figure 35, one of several simple impossible polyhedrons considered by Huffman. Here line 1 is common to surfaces *A* and *B*, line 2 is common to surface *B* and the hidden back surface, and line 3 is common to the hidden surface *C* and surface *A*. On the basis of the same argument as before these three lines must meet at a common point. As the dotted extensions indicate, however, they do not. Hence the figure is impossible. It is strange, Kim observes, that this figure looks so clearly possible in contrast to the Penrose triangle, although the reason for the impossibility of the two figures is the same. Huffman

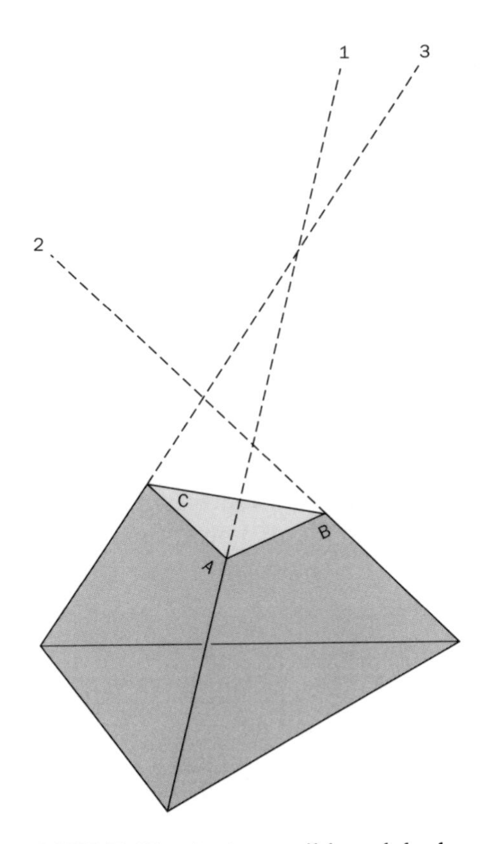

FIGURE 35 An impossible polyhedron

develops a general algorithm, based on directed graphs, for testing all such figures.

The problem of distinguishing possible from impossible polyhedral "windows" has also been investigated, along different lines, by Thaddeus M. Cowan, a psychologist. Basing his analysis on braid theory, Cowan has devised a systematic way of generating and classifying such figures and demonstrating various properties. (See his two papers cited in the bibliography.)

Polyhedral prismatic rings that have square cross sections are easily built from unit cubes. One can rubber-cement together sugar cubes or children's wood blocks, or use the plastic snap-together cubes available from supply houses for mathematical teaching aids. Here is a delightful problem suggested by Kim that makes use of unit cubes.

The prismatic ring in Figure 33 clearly can be constructed with 22 unit cubes as indicated. Our problem is to model a figure with the same properties—one "face" and one "edge"—but to do it with the smallest possible number of cubes.

Each cube must have just two faces joined to other cubes to form the prismatic chain. The light lines, which indicate these joins, are not of course part of the model's edge. The single face of the ring circles the polyhedral torus four times, and it is bounded by a single edge that also circles the ring four times. We make one proviso: no point on the edge may touch any other point on the edge. This is to keep the toroid's hole from being partly or entirely closed.

Figure 36 shows a beautiful wood model of a polyhedral ring, of square cross section, in the shape of a triangle. It was made by Ikuo Sakurai of Tokyo. A section of the model can be twisted to four positions. Down the middle of each face runs a groove inside which a red marble is trapped. Thus the model can be set for any of its four forms. Then by tipping the triangle you can roll the marble once, twice, or four times around the toroid.

Howard P. Lyons, a Toronto accountant, once proposed the following solid prismatic ring. It's outside surface has a square cross section with one twist. An interior hole that circles the ring also has a square cross section with one twist, but the twist goes the other way. In a letter to me, Lyons wonders what properties this bizarre ring has. I take my cue from Mark Twain. He once responded to a man who had asked an unusually complicated question about the speed of a cannonball by writing, "I don't know."

FIGURE 36 A Japanese model of a polyhedral torus adjusted to display a single side and a single edge

ANSWER

The unique 10-cube solution (not counting rotations and reflections) is given in Figure 37.

ADDENDUM

Lawrence H. Peavey, Jr., of Sudbury, Massachusetts, sent me a copy of his U.S. Patent No. 3,010,425 (November 28, 1961) which describes ways in which twisted prismatic rings can be used for what are called "loop scale" devices that measure or control instruments.

Carlo Sequin, at the University of California, Berkeley, was intrigued by Howard Lyons's suggestion about nested

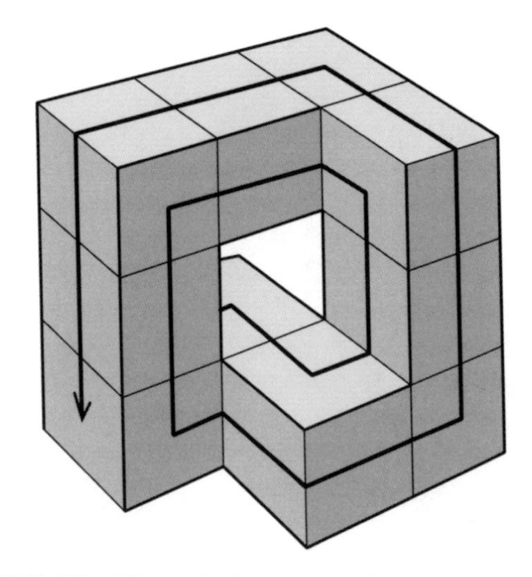

FIGURE 37 The solution to a twisted ring problem

prismatic rings. See the bibliography for his 1980 paper in which he explores the properties of such bizarre structures when the inner and outer surfaces are joined, as in a Klein bottle.

L. Richardson King, at Davidson College in North Carolina, was equally intrigued by my question: "Are there formulas that, given n and k and the method of cutting, will predict the number of twists in the resulting ring or rings?" He sent some fascinating preliminary results which so far have not been published.

When this chapter first appeared as a *Scientific American* column, I cited Charles J. Matthews's 1972 articles, entitled "Some Novel Möbius Strips," as the earliest English reference to the topic that I knew. H. T. McAdams, a scientist at Calspan Corporation, in Buffalo, sent me a copy of his 1948 article (see the bibliography) in which the rings are discussed, perhaps for the first time. Shortly after his article appeared, his table of numbers, relating properties of the rings to Eratosthenes's sieve for finding primes, appeared on the cover of *The Fantopologist,* a science-fiction "fanzine." It is essentially the same as the table in Figure 31. This may have been where Ted Sturgeon got the idea for his 1949 story.

BIBLIOGRAPHY

"The Sieve of Eratosthenes and the Möbius Strip." H. T. McAdams, in *American Mathematical Monthly,* 55, 1948, pages 308–309.

"What Dead Men Tell." Theodore Sturgeon, in *Astounding Science Fiction,* November, 1949. The entire issue was a whimsical joke. In a letter published in November of the preceding year a reader had "reviewed" the issue. The editor, John Campbell, Jr., persuaded most of the writers named in the letter to write stories with the titles that had been mentioned, so that when the issue was published, it made the letter precognitively accurate. Sturgeon's story was reprinted in the collection *Imagination Unlimited,* edited by Everett Bleiler and T. E. Dikty (Berkley Publishing Company, 1959).

"Impossible Objects as Nonsense Sentences." D. A. Huffman, in *Machine Intelligence: Vol. 6,* edited by Bernard Meltzer and Donald Michie. Elsevier, 1971.

"Some Novel Möbius Strips." Charles Joseph Matthews, in *Mathematics Teacher,* 65, 1972, pages 123–126.

"The Theory of Braids and the Analysis of Impossible Figures." Thaddeus M. Cowan, in *Journal of Mathematical Psychology,* 11, 1974, pages 190–212.

"Organizing the Properties of Impossible Figures." Thaddeus M. Cowan, in *Perception,* 6, 1977, pages 41–56.

Triad Illusions and How to Make Them. Harry Turner. Dover, 1978.

"Dulcimer Peg Box: Designer Proposes One-sided Solution." Stanley Hess, in *Fine Woodworking,* Summer, 1978, pages 77–79.

"Twisted Prismatic Klein Bottles." Carlo H. Sequin, in *American Mathematical Monthly,* 87, 1980, pages 269–278.

"Impossible Shapes That Defy the Mind." Dr. Krypton [Paul Hoffman], in *Science Digest,* February, 1984, pages 92–94.

6

The Thirty Color Cubes

Two volumes of *Percy Alexander MacMahon: Collected Papers* have been published. Edited with skill and admiration by George E. Andrews, a mathematician at Pennsylvania State University, the collection is one of a distinguished series of papers of modern mathematicians that is being published by the MIT Press under the general editorship of Gian-Carlo Rota, an MIT mathematician.

We honor MacMahon in this chapter because he was keenly interested in recreational mathematics. His fame rests, however, on more "serious" work. Indeed, he was one of the great pioneers of combinatorics, particularly in the field of number-partition theory. His two-volume magnum opus *Combinatory Analysis* has been reprinted by the Chelsea Publishing Company, but as Andrews points out, that work refers to fewer than a fifth of MacMahon's papers. More than a fourth of his papers appeared after *Combinatory Analysis* was written. It is remarkable how often MacMahon's results are rediscovered by mathematicians who until now have not had easy access to his voluminous writings.

"With his moustache, his British 'Empah' demeanor and worst of all his military background," writes Professor Rota in an introduction to the *Collected Papers,* "MacMahon was hardly the type to be chosen by Central Casting for the role of the Great Mathematician." He was born in 1854 on the island of Malta, the son of a British brigadier. He joined the Royal Artillery in his late teens, served for a time in India and then for many years taught mathematics and physics at the Royal Military Academy. For 14 years before his death in 1929 Major MacMahon was Deputy Warden of the Standards under the Board of Trade.

It is hard to understand why until recently no publisher has reissued MacMahon's long out-of-print *New Mathematical Pastimes,* published by Cambridge University Press in 1921. (It is included in the second volume of *Collected Papers.*) The book deals with tiling theory and repeated patterns, now a lively research topic. MacMahon deals with the subject by way of what he calls a generalized domino. Ordinary dominoes are rectangles with numbered ends. They are employed in a variety of games where they must be placed in chains so that all joined ends have the same number. MacMahon generalized ordinary dominoes to convex polygons that tile the plane. All edges of the polygonal dominoes are labeled with numbers or colors in all possible ways (given the number of labels) to form a complete set of polygons no two of which are alike. (Reflections are included in each set but rotations are not.)

An early column of mine on MacMahon's 24 color squares, representing all the ways of giving an edge one of three colors, is reprinted in my *New Mathematical Diversions from Scientific American.* A later column on MacMahon's 24 color triangles (using four colors for the edges) is in my collection, *Mathematical Magic Show.* MacMahon's *New Mathematical Pastimes* contains a wealth of problems based on these and other sets of edge-colored polygons, including pentagons and hexagons. The book also explains how the edges of such tiles can be altered to produce beautiful periodic patterns of the kind seen in mosques and in M. C. Escher's tessellation pictures of birds and other creatures. Some remarkable recent discoveries about nonperiodic tiling resulted from the work of Hao Wang and others on nonperiodic tiling with MacMahon color squares.

Most of MacMahon's work was centered on symmetric functions, or functions that are unaltered when any two variables are interchanged, for example $abc+a^2+b^2+c^2$. (Interchanging a and c gives $cba+c^2+b^2+a^2$, clearly the same function.) It is easy to see how the polygons in a complete set of MacMahon color polygons are related by a symmetric function. For example, suppose we have a set of 24 color squares representing all the ways of coloring each edge red, blue or green. If we permute the colors any way we like, making, say, all the red edges green, all the green edges blue and all the blue edges red, we end up with exactly the same set of tiles as before. It is this permutation symmetry that underlies the beautiful combinatorial properties of the set.

Color dominoes obviously generalize to solids of three or more dimensions. If each of the n faces of a regular solid is given one of n colors, in how many different ways (not counting rotations but including reflections) can the solid be n-colored? The answer is given by the simple formula $F!/2E$, where F is the number of faces and E is the number of edges.

Of the five Platonic solids, only the cube tiles space, and so it is natural that MacMahon chose to explore sets of "color cubes" and their domino-tiling properties. A cube has six faces and 12 edges. Plugging these values into the formula gives 6!/24, or 30. Thus there are just 30 ways to color a cube with six colors so that all six appear on each cube, with each face a single color. Such a set is known as the 30 color cubes, and MacMahon was the first to investigate its properties. He introduced the set in a lecture given in 1893 and later discussed it briefly in *New Mathematical Pastimes*.

Unfortunately no set of the 30 color cubes can be bought (as far as I know they have never been marketed), but if you will go to the trouble of obtaining 30 wood cubes and coloring their faces correctly (an alternative to painting is to paste colored spots on the faces), you will have a marvelous educational toy. For someone interested in combinatorics, exploring the properties of the 30 cubes can lead into fascinating corners of the subject.

The coloring of the 30 cubes is shown in Figure 38. The cubes are arranged in a six-by-six matrix with one of the diagonals left blank for reasons that will be explained below. The interior square of each figure in the matrix is the top face of a cube. Each cube face is colored and is marked with a corresponding number, and the bottom face of each cube is assigned the missing color and number. Obviously it does not matter what color is assigned to each number.

By labeling the rows with uppercase letters and the columns with lowercase letters we can identify each cube by a pair of letters, one uppercase and one lowercase. Each cube has a mirror-image partner (the cube you would see reflected if you held a cube up to a mirror) that can be quickly located because partners are placed symmetrically with respect to the blank diagonal. Thus the mirror image of cube *Ab* is cube *Ba,* the mirror image of cube *Ec* is cube *Ce* and so on.

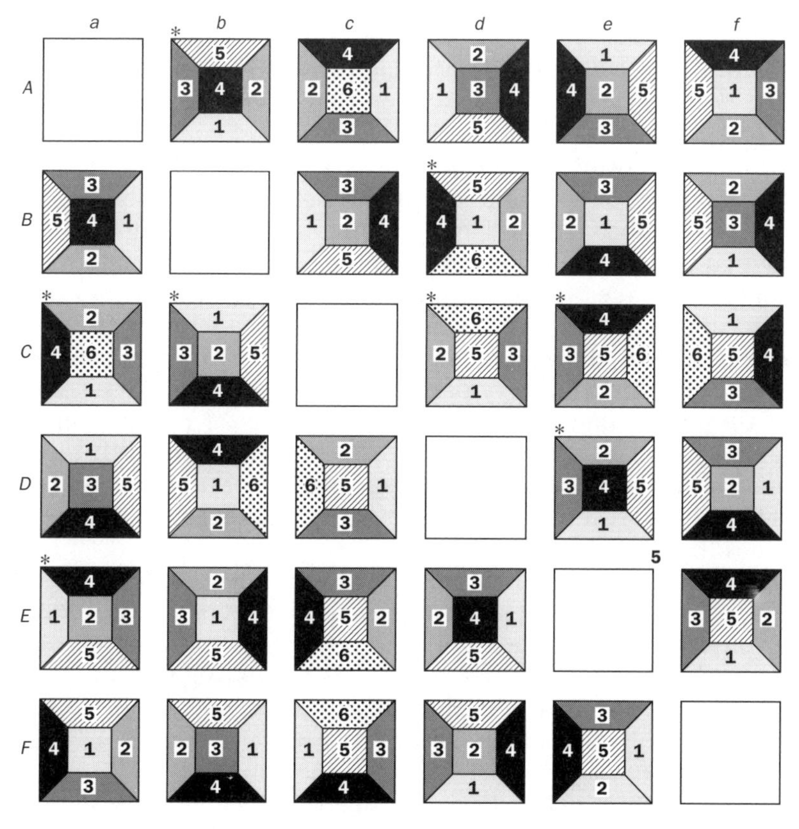

FIGURE 38 John Horton Conway's matrix for the 30 color cubes

If the cubes are scattered on a table, finding the mirror-image partner of a cube can be a chore. Here is a fast procedure for doing it. Suppose cube *X* is red on the top, orange on the bottom, yellow on the front, green on the back, blue on the left and white on the right. You want to find its mirror image. Turn the other 29 cubes so that they are all orange on the bottom. Just five of them will be red on the top. Keep those cubes and discard the others. Rotate the five so that they all have yellow faces toward you. Only one cube will be green on the back, and it is the mirror image of *X*. The procedure works because the mirror-image cube is the only other cube that has the same pairing of colors on opposite faces.

Similar elimination procedures are useful in working on color-cube problems. To speed up such procedures you can arrange a large number of cubes in a row; the entire row can be turned over by applying pressure on the ends. For example, suppose you want all cubes that have red and blue on opposite faces. Turn the 30 cubes red side up and arrange them in several rows. Rotate each row 180 degrees and you can quickly obtain the six cubes that have blue now on the top.

The historic 30-cube puzzle is stated as follows: Select any cube at random and call it the prototype. The task is to find eight cubes among the remaining 29 that will build a two-by-two-by-two model of the prototype. The model must have solid-color faces (each face made up of four cube faces) that correspond to the arrangement of colors on the prototype. Furthermore, the model must meet what we shall call the domino condition. Every pair of touching faces in the interior of the model must be of like color, that is, a red face must abut a red face, a green must abut a green and so on. It turns out that for each prototype there is only one set of eight cubes that will fulfill these requirements, but the cubes will always build the model in two distinct ways.

There is more on this problem and others involving the 30 cubes in the last part of the chapter in *New Mathematical Diversions* cited above and in the references given in the bibliography of that book. My reason for discussing the set again here is to introduce a truly remarkable arrangement of the set, the six-by-six matrix shown in Figure 38, and some unpublished puzzles that make use of the cubes.

This matrix is the discovery of John Horton Conway, now at Princeton University, who also suggested the labeling. The most surprising feature of the matrix is that it instantly provides all the solutions for the historic problem just described. Suppose cube *Df* is chosen as the prototype. To find the eight cubes that model it, first locate its mirror partner *Fd*. The eight cubes are the four that are in the same row and the four that are in the same column as *Fd*, excluding *Fd* itself. The situation is symmetrical. The eight cubes that model *Fd* are the four that are in the same row and the four that are in the same column as *Df*, excluding *Df* itself. This simple procedure applies to each of the 30 cubes.

Other properties of Conway's matrix are only beginning to be explored; perhaps readers can find some new ones. I shall mention only one more property: The matrix also provides instant solutions to a new and more difficult puzzle invented by Conway. The task of the puzzle is to find a set of five cubes with the curious property that if they are turned so that any given color is on the bottom of all five cubes, then the remaining five colors will show on the top. Thus if the cubes are turned so that all have red on the bottom, the other five colors will appear on the top; if the same five cubes are turned so that all have blue on the bottom, again the other five colors will appear on the top, and so on for any color chosen for the bottom.

Each row and each column of Conway's matrix is a set of five cubes that solves this problem. Moreover, the only other sets of five cubes that yield solutions are those obtained by taking one of the 12 sets given by the rows and columns and replacing one or more cubes with a mirror-image cube.

The five cubes in any row or column will also form a one-by-one-by-five prism with the following properties: (1) one side of the prism is all one color, and that color can be any one of the six colors; (2) each of the other three sides displays all five of the other colors; (3) each of the four pairs of touching faces meets the domino condition, and (4) the two ends of the prism are the same color.

As Figure 38 demonstrates, the 30 cubes in Conway's matrix can be oriented so that they meet the domino condition throughout. If we ignore the orientation of individual cubes, their arrangement in the matrix is unique in the following sense. We do not differentiate among (1) different assignments of colors to the numbers, (2) rotations or reflections of the matrix or (3) transformations accomplished by exchanging any pair of columns and then exchanging the corresponding pair of rows to restore the empty diagonal. For example, the matrix formed by switching columns b and f and then switching rows B and F is not considered to be different from the original matrix. (This transformation can be used to put any cube in any specified position.)

An old puzzle involving four color cubes was a big seller in 1968 when Parker Brothers marketed it under the trade name Instant Insanity. It is not germane to our topic because the cubes of the puzzle are not members of the set of 30. (Duplicate colors appear on each cube.) Many problems more interesting than Instant Insanity can be created, however, by selecting subsets from the 30. MacMahon himself sold an eight-cube puzzle to the London company R. Journet, which marketed it around the turn of the century as the Mayblox puzzle. Its eight cubes were simply one of the 30 sets of eight that model a prototype and meet the domino condition. Purchasers were not told which prototype to model, however, and without this information the task is more difficult.

If the domino condition is discarded and one simply tries to build a larger cube that models a prototype, much more difficult puzzles can be invented. One of the best is the work of Eric Cross of Ireland. In 1970 it was sold in the U.S. by Austin Enterprises of Akron, Ohio, under the trade name Eight Blocks to Madness. The eight cubes in this subset are the ones shown marked with an asterisk in Conway's matrix. Make a set (or select it from your set of 30) and see if you can build a model of one of the remaining cubes. Remember, you do not have to meet the domino condition. It is only necessary to model the outside colors of one of the 30 cubes. Only one such cube can be modeled, but it can be done in two ways.

Here is another puzzle involving the same set of eight cubes: Form a larger cube with four distinct colors on each face and each of the six colors represented just four times on the outside of the cube. Once again the solution is not required to fulfill the interior domino condition.

One of the pleasures of playing with the 30 cubes is that one can invent new problems and confront the challenge of either finding solutions or proving impossibility. For example, is it possible to find a set of six cubes that form a one-by-one-by-six prism for which each of the four sides shows all six colors, all pairs of touching faces as well as the two ends match in color and all six colors are represented by the matching pairs? The answer is yes. Is it possible to divide the 30 cubes into five distinct sets of six, each of which solves this problem? I do not know.

Take any eight-cube model that solves the classic problem and join it to its mirror-image model with the like-color faces of the two models touching. The result is a two-by-two-by-four brick with a different color on each of its four larger faces and a fifth color on the other two faces. The brick will of course meet the domino condition throughout. It is easy to think of scores of similar problems for bricks of other measurements, but I know of no reported results.

In 1956 Paul B. Johnson showed that when the domino proviso is dropped, there are 144,500 ways to build an eight-cube model of a prototype that is not one of the eight cubes. Of this number, 67,260 ways build with different sets of eight cubes. Johnson also proved that any set of eight cubes that models a prototype (not including the prototype) will do so in two, four, eight or 16 ways. Conway, working with a system of directed graphs, obtained the last result independently and also showed that no set of eight will build a prototype in just one way even if the prototype is among them. If a set builds a prototype at all, it will do so in at least two ways. If a set of eight cubes can model a prototype (not including the prototype) in 16 ways, Conway proved that two of the ways will satisfy the domino condition.

Conway has obtained many other results on the 30 cubes that have not been published. Of particular interest is a quick method for finding in his matrix the prototype, if there is one, that can be modeled (without meeting the domino proviso) by any given set of eight cubes. Here are some research questions, some of which Conway has solved, about such models:

> Is it possible to select three disjoint sets of eight cubes from the 30 that will build three models of the same prototype?
>
> Is it possible to select three disjoint sets of eight cubes that will model three distinct cubes?
>
> What is the largest set of different cubes for which there is only one subset of eight cubes that will model a cube?

Major MacMahon's set of 30 cubes, already a classic source of recreational mathematics, surely conceals a wealth of surprises yet to be discovered.

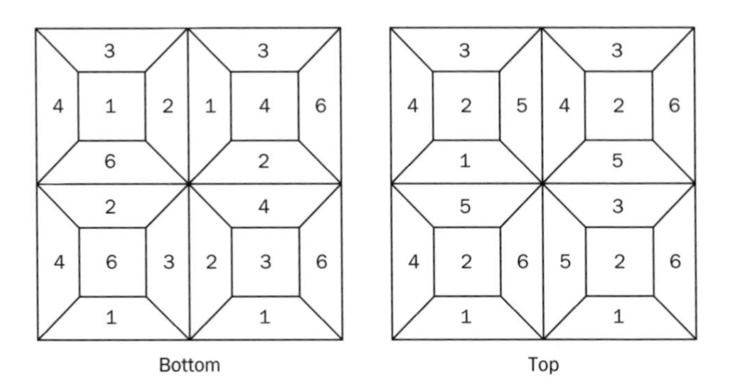

FIGURE 39 Solution to the "Eight Blocks to Madness"

ANSWERS

The problem was to use eight specified color cubes to build a larger cube with each face a solid color and no two faces the same color. The top and bottom layers of one solution are shown in Figure 39. The second solution is obtained by shifting the cubes in the top layer so that each cube moves counterclockwise to the adjacent position without changing its orientation. The prototype of the model is cube *Fc* in the matrix of 30 color cubes.

The second problem was to use the same eight cubes to build a larger cube that has four different colors on each face, with each color represented four times in all. Two solutions (I do not know if there are others) are easily obtained from the two solutions to the previous problem. Simply "triple cut" each cube by exchanging the left and right slabs, the front and back slabs and the top and bottom slabs. The three cuts can be made in any order.

Two solutions are not, of course, "different" if one can be obtained from the other by rotations, reflections or permuting the colors.

ADDENDUM

I said I did not know whether the 30 cubes can be divided into five sets of six each, such that each set is a row showing six different faces on each of its four sides, with end colors matching, and all touching faces matching. Zoltan Perjés, of the General Institute of Physics, Budapest, was the only reader who solved this problem. A solution is shown in Figure 40. Nontrivially different sets, he pointed out, also solve the problem.

I regret not having space to discuss the long, fascinating letters from Michael G. Harman, Peter Cameron, D. P. Lau-

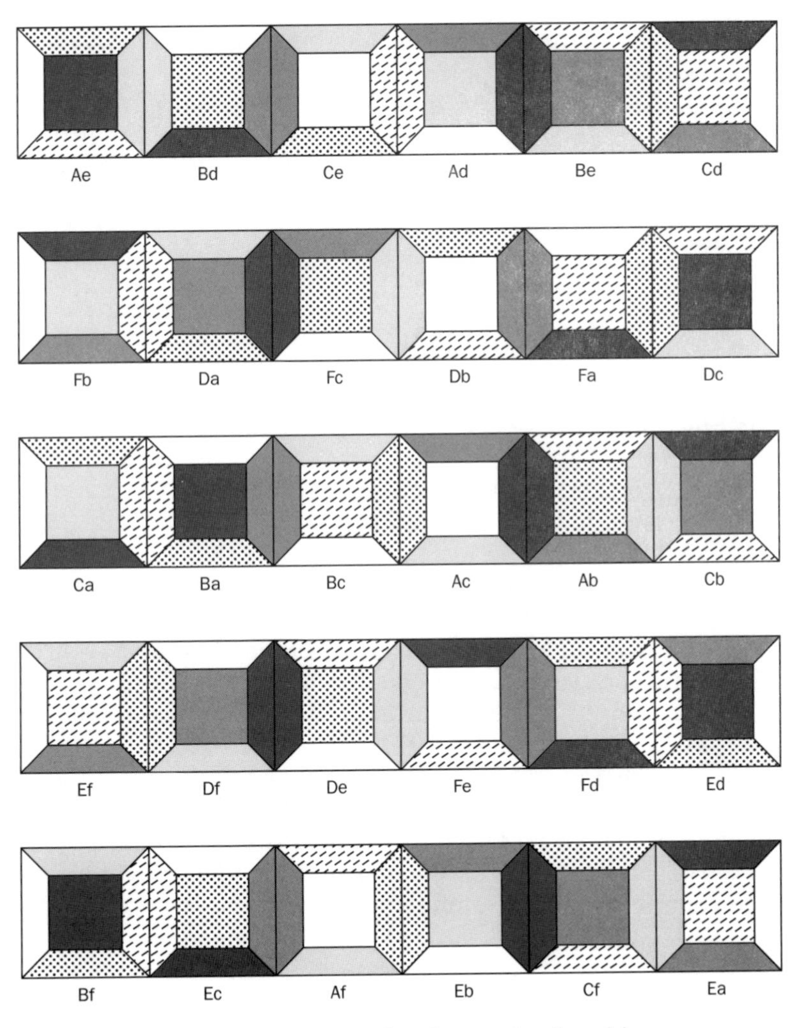

FIGURE 40 Solution to a previously unsolved problem

rie, G. J. Westerink, Zoltan Perjés, and others who sent material about the cubes, and called attention to how the cubes relate to significant problems in graph theory and combinatorial geometry.

Harry Sonneborn III found it helpful to dispense with diagrams, and designate each cube by six digits, taking in order the left side, right side, top, bottom, far side, near side. Thus the cube at the lower left corner of Conway's matrix would be coded by 421653. A doublet at either end can be moved to the middle without altering the cube, but if a doublet is put between the other two it changes the cube to its mirror image. Transposing the numbers of any doublet also changes the cube to its mirror reflection.

Because there are 15 different ways (ignoring rotations) to designate a cube by three doublets, these 15 and their mirror forms prove there are just 30 different cubes. Sonneborn went on to show how many of the problems I posed could be solved simply by manipulating the digits.

BIBLIOGRAPHY

Das Spiele der 30 Bunten Würfel. Ferdinand Winter. Leipzig, 1934.

"Colored Polyhedra: A Permutation Problem." Clarence R. Perisho, in *Mathematics Teacher,* 53, 1960, pages 253–255.

The Cube Made Interesting. Aniela Ehrenfeucht. Pergamon, 1964, pages 46–66.

New Mathematical Diversions from Scientific American. Martin Gardner. Simon and Schuster, 1966, Chapter 16.

"The 23 Colored Cubes." Norman T. Gridgeman, in *Mathematics Magazine,* 44, 1971, pages 243–252.

"Coloring the Faces of a Cube." E. H. Lockwood, in *Mathematical Gazette,* 61, 1977, pages 179–182.

"A Notation for MacMahon's Coloured Cubes." Dirk P. Laurie, National Research Institute for Mathematical Sciences, Pretoria, South Africa, Technical Report, TWISK 40, August, 1978.

On 8-cube (2 × 2 × 2) Puzzles

"Stacking Colored Cubes." Paul B. Johnson, in *American Mathematical Monthly,* 63, 1956, pages 115–124.

"Colour Cube Problem." W. R. Rouse Ball, in *Mathematical Recreations and Essays* (revised), Macmillan, 1960, pages 112–114.

"The Mayblox Problem." Margaret A. Farrell, in *Journal of Recreational Mathematics,* 2, 1969, pages 51–56.

" 'Eight Blocks to Madness'—A Logical Solution." Steven J. Kahan, in *Mathematics Magazine,* 45, 1972, pages 57–65.

"More Progress to Madness Via 'Eight Blocks.' " Andrew Sobczyk, in *Mathematics Magazine,* 47, 1974, pages 115–124.

On Instant Insanity

Puzzles and Paradoxes. T. H. O'Beirne. Oxford University Press, 1965, Chapter 7.

"Crazy Cubes." Paul Wehl, in *Popular Science,* 193, 1968, pages 132–133.

"A Note on 'Instant Insanity.' " T. A. Brown, in *Mathematics Magazine,* 41, 1968, pages 167–169.

"Solving 'Instant Insanity.' " Robert E. Levine, in *Journal of Recreational Mathematics,* 2, 1969, pages 189–191.

"An Improved Solution to 'Instant Insanity.' " B. L. Schwartz, in *Mathematics Magazine,* 43, 1970, pages 20–23.

"A Diagrammatic Solution to 'Instant Insanity' Problem." A. P. Grecos, in *Mathematics Magazine,* 44, 1971, pages 119–124.

"A Cure for 'Instant Insanity.' " Edward J. Wegman, in *Pi Mu Epsilon Journal,* 5, 1971, pages 221–223.

" 'Instant Insanity,' that Ubiquitous Baffler." Dewey C. Duncan in *Mathematics Teacher,* 65, 1972, pages 131–135.

"A $2 \times 2 \times 1$ Solution to 'Instant Insanity.' " Kay P. Litchfield, in *Pi Mu Epsilon Journal,* 5, 1972, pages 334–337.

" 'Instant Insanity'—A Significant Puzzle for the Classroom." Joseph A. Troccolo, in *Mathematics Teacher,* 68, 1975, pages 315–319.

"Generalized 'Instant Insanity' and Polynomial Completeness." Edward Robertson and Ian Munro, in *Proceedings of the 1975 Conference on Information Science and Systems,* April 2–4, 1975.

"NP-Completeness, Puzzles, and Games." Edward Robertson and Ian Munro, in *Utilitas Math.*, 13, 1978, pages 99–116. The authors show in this and the previous paper that a generalized Instant Insanity problem is NP-complete.

7

Egyptian Fractions

Long before the Christian Era, Chinese mathematicians had a surprisingly sophisticated comprehension of fractions. They accepted any whole number as a numerator or denominator, and had excellent rules for adding, subtracting, multiplying and dividing fractions. As is common practice today, the Chinese preferred to work with proper fractions, or fractions that have the numerator smaller than the denominator, a feature reflected in their calling the numerator *tzu* (son) and the denominator *mu* (mother).

The ancient Egyptians, however, had a peculiarly hobbled approach to fractions. They understood rational fractions with numerators greater than 1 well enough but apparently could not deal with them as single numbers. With the sole exception of 2/3, for which there was a special hieroglyph, they had symbols only for unit fractions, that is, fractions that are the reciprocals of positive integers, with 1 above the line and any positive integer below.

To manipulate fractions with numerators higher than 1 the Egyptians expressed such fractions as sums of distinct unit fractions. For example, instead of writing 5/6 they wrote 1/2 + 1/3. They devised rules for carrying out all the necessary arithmetical operations on expressions of this type. In certain cases, particularly those involving addition, there are advantages to working with fractions in the expanded form, but in general the Chinese methods for handling fractions are far superior.

Most of what is known about Egyptian fractions is derived from the information given in a famous document, now known as the Rhind papyrus, which was inscribed in about 1700 B.C. It was bought in Luxor in 1858 by A. Henry Rhind, a Scottish antiquary, and is now owned by the British Museum. The papyrus, a kind of calculator's handbook, opens with a table in which every fraction of the form $2/b$ is expressed as a sum of distinct unit fractions arranged in decreasing order of size, where b is equal to all the odd integers from 5 through 101.

Did the Egyptians have a systematic method for expanding proper fractions in this way? A number of scholars have speculated about the possibility but it seems most probably that they did not, because the expansions given in the Rhind papyrus are not always the "best." There are of course many different ways to define "best." The most obvious way is to call an expansion best if it minimizes the number of terms. Another type of best expansion is one that minimizes the largest denominator in the series. For example, the expansion of 3/7 as 1/4 + 1/7 + 1/28 has the smallest possible number of terms, but the expansion as 1/6 + 1/7 + 1/14 + 1/21 has the smallest possible value of the largest denominator. If both the number of terms and the largest denominator can be minimized in the same expression, so much the better. (I shall not consider other types of best expansions such as those that minimize the sum of all the denominators.)

Even stranger than the preference of the Egyptians for such a cumbersome system is that the Greeks adopted it. In fact, the system was widely used in Europe until well into the 17th century! Even the great Archimedes calculated with what are now called Egyptian fractions. The term has come to refer to any expression of a rational number as a sum of distinct unit fractions, traditionally arranged in decreasing order of size. Modern fractions, in which any positive integer can be above or below the line, are derived from Hindu mathematics and were not widely adopted until the 18th century. Some have maintained that the long preoccupation with unit fractions was a cultural bias that delayed progress in mathematics as much as the Roman system of writing numerals did.

Investigating the properties of Egyptian fractions is now a small but challenging task in number theory. There are many deep unsolved problems in this area, but there are also many problems well within the reach of any clever novice that have much in common with certain recreational puzzles. For example, consider the old Arabian brainteaser about a man whose will specified that his 11 horses be divided so that his eldest son would get 1/2, his middle son would get 1/4 and his youngest son would get 1/6. When he died, his lawyers were puzzled about how to carry out these eccentric instructions. After all, horses are of little value when sliced into fractional parts. A relative, hearing of the problem, solved it by lending the heirs his own horse. The 12 animals were then easily divided according to the formula in the will, with the three sons respectively getting six, three and two. One horse was then left over, and so the relative got his horse back!

The puzzle has appeared in many different forms, and of course it can be generalized to deal with larger numbers of sons and larger numbers of horses that are borrowed and then returned. If we stick to the story's traditional form involving three sons and one borrowed horse, an interesting question arises. How many variations are possible in the number of horses to be divided and the set of three fractions for dividing them specified in the father's will? One might guess that there would be an infinite number, but there are only seven. They are the seven solutions of the Diophantine equation $n/(n+1)=1/a+1/b+1/c$, where a, b and c are positive, distinct integers, a is less than b, b is less than c and $n+1$ is the least common multiple of a, b and c.

It is easy to show that a must be equal to 2. If a is greater than 2, then the lowest possible least common multiple for a, b and c is 12, obtained when a equals 3, b equals 4 and c equals 6. Therefore $n/(n+1)$ must be at least 11/12. But the sum $1/3+1/4+1/5$ equals 47/60, which is less than 11/12, and if the denominators are raised, the sum is even smaller. Hence

a is not greater than 2, and so a equals 2. A similar argument shows that b must be either 3 or 4, and with that information it is not hard to determine all the possible values for c. The chart in Figure 41 gives the value of n (the original number of horses to be divided) and the denominators of the three unit fractions for each of the seven possible variations of the puzzle.

It is obvious that any proper fraction can be expressed as the sum of unit fractions if a repetition of terms is allowed. For example, 3/7 equals 1/7 + 1/7 + 1/7. It is not obvious, however, that every proper fraction can be expressed as the sum of unit fractions even if a repetition is forbidden. One proof of this fact is the existence of a famous algorithm for writing any proper fraction as the sum of a finite number of distinct Egyptian fractions. The algorithm was first published in 1202 by Leonardo of Pisa, better known as Fibonacci, in his influential book on arithmetic *Liber abaci.* Fibonacci preferred to work with unit fractions, and his book contains tables for converting proper fractions to Egyptian sums. His algorithm for converting a proper fraction to the sum of a finite number of distinct Egyptian fractions is given without any proof that it always works. The eminent British mathematician J. J. Sylvester rediscovered the algorithm, and in 1880 he published the first proof that it does always work.

	n	*a*	*b*	*c*
1	7	2	4	8
2	11	2	4	6
3	11	2	3	12
4	17	2	3	9
5	19	2	4	5
6	23	2	3	8
7	41	2	3	7

FIGURE 41 All the variations of an old Arabian puzzle

Fibonacci's method is simple. Call the proper fraction a/b. The first term of the expansion is the largest unit fraction not greater than a/b. Now subtract the unit fraction from a/b to obtain another proper fraction. The second term of the expansion is the largest unit fraction not greater than this remainder. Continue in this manner, each time putting down the largest usable unit fraction as the next term in the expansion and then subtracting and repeating the process with the remainder. It is clear that the fractions obtained in this way will grow steadily smaller. It can be proved that the process always terminates. Hence the algorithm always works. (It is also possible to express any irrational number as the sum of an infinite series of distinct unit fractions, but that is too far removed from our topic.) In today's vernacular, Fibonacci's method is known as a greedy algorithm, because at each step in the process the largest fraction possible is chosen.

Although the greedy algorithm will express any proper fraction in the Egyptian manner, it does not always give the best expansion in either of our two senses of the word. When the algorithm is applied to fractions of the form $1/b$, however, it does generate the best two-term expansion of $1/b$ in both senses. A little algebraic doodling will convince you that applying the algorithm is the same as replacing $1/b$ with $1/(b+1)+1/[b(b+1)]$. Thus $1/2$ equals $1/3+1/6$; $1/3$ equals $1/4+1/12$; $1/4$ equals $1/5+1/20$, and so on.

The formula given above for a two-term Egyptian expansion of $1/b$ also serves to prove that any proper fraction can be expressed in the Egyptian manner in infinitely many ways. Consider the expansion $2/3=1/2+1/6$. By applying the formula to the last term $1/6$ we obtain a new expansion: $2/3=1/2+1/7+1/42$. If we repeat the procedure with $1/42$, we obtain $2/3=1/2+1/7+1/43+1/1{,}806$. In this way the expansion of $2/3$ as a series of unit fractions can be continued indefinitely.

The same formula underlies an algorithm called the splitting method, which, like Fibonacci's greedy algorithm, is guaranteed to generate a finite Egyptian series for any proper fraction. There are many other algorithms that serve the same purpose, each with its own advantages and defects. Some algorithms minimize the number of terms and others minimize the largest denominator, but all of them, including the greedy algorithm, are inefficient and difficult to apply to fractions with large denominators and numerators.

When the greedy algorithm is applied to a proper fraction a/b, it always generates a series of Egyptian fractions with a number of terms no greater than a. Thus when it is applied to a proper fraction of the form $2/b$, it generates an expression with one or two terms: 2/4 equals 1/2; 2/5 equals $1/3+1/15$; 2/6 equals 1/3; 2/7 equals $1/4+1/28$, and so on. At each step the algorithm chooses the largest unit fraction that is smaller than the remainder, but since each step is unaffected by preceding or succeeding steps, the procedure can easily miss an expression with fewer terms when it is applied to a proper fraction with a numerator of 3 or higher. It also tends to generate terms with denominators much larger than necessary.

Michael N. Bleicher, in his section on Egyptian fractions in the book *Excursions into Mathematics* gives several horrendous examples of how miserably the greedy algorithm can fail to generate the best expansion in either of our two senses of the word. For example, when the algorithm is applied to 5/121, it generates the series

$$\left[\frac{1}{25}, \frac{1}{757}, \frac{1}{763309}, \frac{1}{873960180913}, \frac{1}{1527612795642093418846225}\right].$$

Bleicher compares that with the expansion $5/121=1/25+1/759+1/208{,}725$. There is no way to express 5/121 with fewer than three terms, but Bleicher does not know whether or not a three-term expansion can have a largest denominator smaller than 208,725.

For proper fractions of the form $3/b$ the greedy algorithm is guaranteed to generate an expression of three or fewer terms, and if the fraction has the form $4/b$, it is guaranteed to generate an expression of four or fewer terms. The outstanding unsolved question of Egyptian fractions concerns the case $4/b$: Can a proper fraction $4/b$ always be expressed with three or fewer terms? In other words, can the Diophantine equation $4/n=1/a+1/b+1/c$ always be solved in positive integers for any integral value of n greater than 4?

Paul Erdös and E. G. Straus have conjectured that the equation can always be solved. Their conjecture has been verified to extremely large values of n, but it has not been proved. Similarly, Waclaw Sierpinski has conjectured that all proper fractions of the form $5/b$ can be similarly expressed with no more than three terms. He has also conjectured that for any given integer k and a variable with integral values b there is a value of b greater than k such that for all larger values of b the proper fraction k/b can always be expressed with no more than three terms.

So far we have considered only proper fractions. What about improper fractions such as 2/1 and 7/3? These fractions too can always be expressed by a finite Egyptian series in infinitely many ways. To generate such an expression we start with terms in the harmonic series $1/1 + 1/2 + 1/3 + 1/4 + \ldots$, because they are the largest unit fractions we can use. It is well known that the harmonic series does not converge. In other words, a partial sum (the sum of the first n terms for some n) will exceed any integer we name. The procedure for generating a series of Egyptian fractions equal to an improper fraction consists in using the harmonic series as far as possible and then adding more unit fractions to express whatever fractional part of the desired total remains. The harmonic series diverges with distressing and increasing slowness, however, and for this reason even small improper fractions demand enormously long Egyptian expressions. For example, expanding 10/1 requires more than 20,000 unit fractions from the harmonic series.

Recently some curious results concerning Egyptian fractions have been obtained. In 1964 Ronald L. Graham of Bell Laboratories studied the question of what rational fractions can be expressed by Egyptian fractions in which all the denominators are squares. He solved the problem completely, and he also solved the more general problem of determining which rational fractions can be expressed with Egyptian fractions in which all the denominators are powers higher than 2. In the same year Graham also showed that if there is a set of numbers that includes all prime numbers greater than some number and all squares greater than some (possibly different) number, then any rational fraction has an Egyptian expansion that draws all its denominators from the set.

Particularly difficult problems arise when the denominators of Egyptian fractions are limited to the odd whole numbers. It is easy to see that the sum of a series of such fractions cannot be a fraction with an even denominator. It has been shown that every rational fraction with an odd denominator can be expressed as the sum of a finite series of distinct Egyptian fractions, all with odd denominators, for example $2/3 = 1/3 + 1/5 + 1/9 + 1/45$; $2/5 = 1/3 + 1/15$, and $2/7 = 1/7 + 1/9 + 1/35 + 1/315$. There are inefficient algorithms for finding such expressions, but no one has yet proved that Fibonacci's greedy algorithm always terminates when it is applied to this task, even for proper fractions. When denominators may be either even or odd, the greedy algorithm generates increasingly small unit fractions, and it can be proved that the series must terminate. If, however, all denominators are odd, it is not known if the greedy algorithm always terminates.

The expansion of 1 into the smallest number of Egyptian proper fractions with all odd denominators was not found until 1971. (The expression 1/1 is not allowed.) It turns out that there are five solutions to this problem, each with nine terms. The expansion with the smallest largest denominator is $1 = 1/3 + 1/5 + 1/7 + 1/9 + 1/11 + 1/15 + 1/35 + 1/45 + 1/231$. All five solutions start with the reciprocals of 3, 5, 7, 9, 11 and 15. The other four solutions continue with the reciprocals of 21, 135 and 10,395; 21, 165 and 693; 21, 231 and 315, and 33, 45 and 385. What Egyptian expansion for 1 with all odd denominators has the smallest largest denominator? The only answer is the 11-term series $1 = 1/3 + 1/5 + 1/7 + 1/9 + 1/11 + 1/33 + 1/35 + 1/45 + 1/55 + 1/77 + 1/105$.

By adding 1/1 to each of the series given above we get the best odd-denominator Egyptian expressions for 2. When 1/1 is not allowed, however, I do not know what the best expressions for 2 are (in either sense of the term) or even whether such expressions have been found.

Here are four easy Egyptian fraction problems.

1. Express 1 as the sum of three distinct unit fractions.

2. Express 67/120 as the sum of the fewest possible Egyptian fractions, with the smallest largest denominator for that number of terms.

3. The fraction 8/11 is the "smallest" proper fraction that cannot be expressed with fewer than four Egyptian fractions, in the sense that the sum of its numerator and denominator is minimized. Find a four-term expression for the fraction.

4. When the greedy algorithm is applied to a proper fraction of the form $3/b$, what is the smallest value for b such that the algorithm produces an expansion of three terms and that $3/b$ can be expressed as the sum of two distinct unit fractions?

ANSWERS

1. The only solution is the expansion $1 = 1/2 + 1/3 + 1/6$.

2. The expansion is $67/120 = 1/3 + 1/8 + 1/10$.

3. The expansion is $8/11 = 1/2 + 1/6 + 1/22 + 1/66$.

4. The smallest value for b is 25. The greedy algorithm gives the expansion $3/25 = 1/9 + 1/113 + 1/25{,}425$, but there is also the expansion $3/25 = 1/10 + 1/50$.

I reported a question by Michael Bleicher: Can 5/121 be expanded to three unit fractions with a smaller largest denominator than 208,725? The answer is yes. Readers too numerous to mention found the "best" solution: $5/121 = 1/33 + 1/121 + 1/363$.

I am indebted to William Gosper for providing the Egyptian fraction series for 5/121 that is generated by the greedy algorithm. It corrects the third and fifth terms as they appear in Bleicher's contribution to *Excursions in Mathematics*.

ADDENDUM

I said I did not know of the minimum expansion for 2 in odd Egyptian fractions. William Friedmann sent me one with 86 terms, and 6,195 as the largest denominator, and an 89-term expression with 1,765 as the largest denominator. He did not know if either of these are "best."

BIBLIOGRAPHY

"Unit Fractions." Julia Adkins in *Enrichment Mathematics for the Grades, 27th Yearbook,* Chapter 15. National Council of Teachers of Mathematics, 1963.

"Egyptian Fractions." Anatole Beck, Michael N. Bleicher, and Donald W. Crowe, in *Excursions Into Mathematics,* Chapter 6, Section 7. Worth, 1969.

Mathematics in the Time of the Pharaohs. Richard J. Gillings. MIT Press, 1972.

"Egyptian Fractions." Richard K. Guy, in *Unsolved Problems in Number Theory,* Section D11. Springer-Verlag, 1980. Includes four pages of references.

Minimal Sculpture

"Its chief merit is its Simplicity—a Simplicity so pure, so profound, in a word, so *simple,* that no other word will fitly describe it."

—LEWIS CARROLL, *The New Belfry of Christ Church, Oxford*

Modern art, particularly in the U.S., is in such a disheveled state that almost any kind of art—good or bad, traditional or avantgarde, serious or put-on—gets displayed, praised, condemned and even bought. In painting, for reasons over which critics wrangle, there is a strong movement toward realism, but in sculpture most of the movement seems to be in the opposite direction. Not that realism has not invaded the three-dimensional art world as well! Some 15 years ago George Segal started making his plaster-cast models of human figures, at first painted all white and later in bright colors. Now Duane E. Hanson has carried realistic sculpture to its ultimate by creating life-size waxworks of men and women that are the 3-space analogues of color photographs. The main trend, however, at least with respect to the outdoor sculpture found in parks and in front of buildings, has been in the abstract direction of minimal art.

Minimal sculpture is sculpture reduced to extremely simple, nonobjective forms. Junkyard art may be minimal in terms of cost of materials, but its form is usually quite complex. A piece of driftwood art may be minimal in terms of the artist's efforts, but it also is much too complicated to be called minimal. True minimal art—the geometric shapes one now sees in public places—suggests, as art critic Hilton Kramer wrote in *The New York Times,* "the atmosphere of the assembly line, the engineering laboratory, the drafting table and the plastics factory." To Kramer's list I should like to add: illustrations in books on mathematics, in particular books on recreational solid geometry. In this chapter we take a look at some areas where minimal sculpture and geometrical play overlap.

Let us begin with the minimal technique that consists in cutting metal sheets, folding them and perhaps painting them. Picasso's rust-colored 50-foot work that stands in the plaza of Chicago's Civic Center is a striking example of the genre. It was first modeled as a folded cardboard cutout, but it is not really minimal because it has two pupils (although only one eye) and what seem to be two nostril holes (see Figure 42). Picasso's *Bust of Sylvie,* standing 30 feet high on Bleecker Street in New York, is more obviously a woman's face, although it too was modeled as a cutout. Completely nonobjective realizations of the technique in many colors and varieties have been created by minimalists all over the world.

Figure 43 shows a proposal for a public monument of the same type. Study it carefully. Is it possible to construct a model of this form by taking an ordinary file card, snipping it with scissors and folding it, or is it necessary to glue it along certain edges? I do not know the origin of this marvelous new test of one's ability to think in 3-space. I first heard of it from Kim Iles, who teaches forestry at the University of British Columbia. He in turn heard it from a visiting Russian professor of forestry, who had seen it in an entrance examination for the school of architecture at the University of Leningrad.

The form has come to be called a hypercard. Magicians have learned of it and have made it the basis for a number of magic tricks. The hypercard is also the basis for an amusing party game. Place a large model of the card in the center of the floor. The players may view it from any angle but are not allowed to touch it. Each player is given one file card and a

FIGURE 42 Chicago's Picasso. W. B. Finch/Art Resource, N.Y. Copyright © 1991 ARS N.Y./SPADEM.

pair of scissors, and a prize goes to the first player who comes up with a replica of the model. It is surprising how many people decide the task is impossible.

FIGURE 43 Design for a minimum-sculpture monument

Another popular minimal-sculpture technique consists in simply building a large model of a polyhedron. Of recent works of this type one of the best-known is *Cigarette,* a huge, twisted black polyhedral prism designed by minimalist Tony Smith of South Orange, N.J. There is a cartoon by David Levine (*New York Review of Books,* September 26, 1968) that shows the Mary of Michelangelo's *Pietà* holding on her lap the limp form of a polyhedron that looks suspiciously like Smith's *Cigarette.*

Some minimalists like to reproduce an ordinary cube, although their models are usually colored with a spray gun and tipped at an angle so no one will suppose they are merely the pedestal of a work still to come. The best known example is Isamu Noguchi's red cube with a hole going through it, and balanced on one corner at 140 Broadway, in Manhattan (see Figure 44).

Born in Los Angeles in 1904, of a Japanese father and Irish-American mother, Noguchi is one of the most famous of

FIGURE 44 Noguchi's "Red cube," his most famous work, at 140 Broadway, in Manhattan. Courtesy of the Isamu Noguchi Foundation, Inc.

minimalist sculptors. A photograph of Noguchi's cube provides the frontispiece of *Geometry,* a textbook by Christian Hirsch and four others (Scott, Foresman, 1978), and the book's red covers, both front and back, are photographs taken through the cube's hole.

Sticking cubes together in various ways is another favorite ploy of minimalists. When the cubes are joined at their face, the result is known to mathematicians as a polycube. I have often thought Piet Hein's Soma cube, which consists of six tetracubes and one tricube, would make an entertaining piece of outdoor minimal sculpture. Each month its overall structure could be altered, although the pieces would have to be locked together in some way to prevent theft or vandalism. (For a discussion of the Soma cube and other polycube recreations, see Chapter 3 of my *Knotted Doughnuts and Other Mathematical Entertainments,* W. H. Freeman, 1986).

A polycube consisting of eight cubes arranged in a 1-by-2-by-4 rectangular parallelepiped is known as the canonical brick. As far as I know the first sculptor to use canonical bricks in a work of art was the New York minimalist Carl Andre. In 1976 the Tate Gallery in London displayed a work by Andre that consisted of 120 ordinary bricks (not quite canonical, but close to it) packed into a rectangular parallelepiped two bricks high, six wide and ten long. The bricks had been shipped to the Tate by Andre with "directions" for their assembly. Most viewers considered the work nothing more than a "pile of bricks," and London newspapers had a field day when it was disclosed that the Tate had paid Andre $12,000 for it. John Russell of *The New York Times* defended the work for its "order, resolution and . . . absolute simplicity," for its "clarity of intention" and for the "frank and unambiguous way in which the materials are assembled." An Andre, said Russell, "just lies there and minds it own business."

To mathematicians, however, Andre's pile of bricks minds decidedly dull business. They can think of all kinds of ways of packing polycubes that are just as pleasing aesthetically and have the added merit of being interesting. For example, consider the three-dimensional form of the flat *Y* pentomino. The rectangle of smallest area (5 by 10) that can be packed with *Y* pentominoes is shown at the left in Figure 45. If the pentomino is given a unit thickness, so that it becomes a solid of five joined cubes, it is called the *Y* pentacube. What rectangular boxes can be fully packed (without holes) using *Y* pentacubes?

To be fully packable a box must of course have a volume in unit cubes that is a multiple of 5. No fully packable box of volume $5p$ exists, where the number of *Y* pentacubes p is a prime. In 1970 C. J. Bouwkamp and David A. Klarner reported on the results of a computer program that found all the boxes that can be fully packed with 25 or fewer *Y* pentacubes. The smallest is the 1-by-5-by-10 box. One of its four possible packings is shown at the left in Figure 45. There are

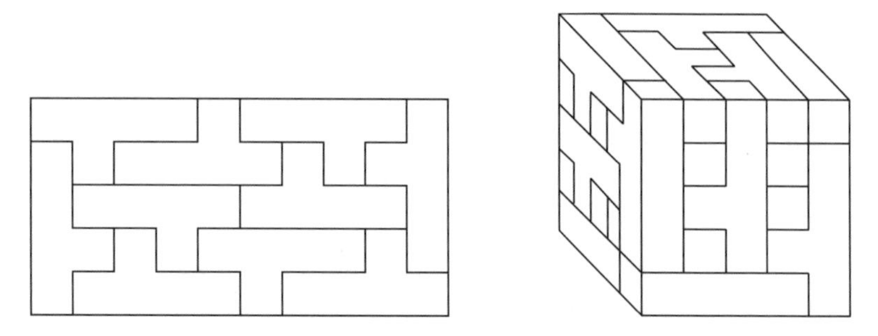

FIGURE 45 *Y* pentominoes (left) and *Y* pentacubes (right)

three other boxes with the same volume, but none is packable.

The smallest cubical box, and the only box of volume 125, that can be fully packed with *Y*'s is the 5-by-5-by-5 cube. If the reader will take the trouble to make 25 *Y* pentacubes, he will find assembling them into a cube is a splendid puzzle. The solution partly given at the right in Figure 45 is one Klarner found by hand before the computer program printed hundreds of other solutions. If Andre's Tate Gallery work had been a cube composed of 25 *Y* pentacubes instead of a rectangular parallelepiped of 120 octacubes, it would have required a more detailed set of assembly instructions, but at least it would have intrigued mathematicians.

Lewis Carroll is among those who have felt that an ordinary cube is too minimal to have much aesthetic value, and no one has been funnier in the written criticism of such art. In 1872 a new belfry was designed to house the bells that had been removed from the cathedral of Christ Church, Oxford, where Charles L. Dodgson taught mathematics. The belfry, placed over an elegant staircase leading to the hall at a corner of the Great Quadrangle, was nothing more than a simple wood cube. The design so annoyed Dodgson that he privately published a monograph on the subject titled *The New Belfry of Christ Church, Oxford.* The title is followed by a line from Keats: "A thing of beauty is a joy forever." Below the quotation Dodgson drew a picture of a square and captioned it: "East view of the new Belfry, Ch. Ch., as seen from the Meadow (see Figure 46)."

Dodgson opens his monograph with a note of etymology. The word "belfry," he writes, is from the French, *bel,* meaning

THE NEW BELFRY

OF

CHRIST CHURCH, OXFORD.

A MONOGRAPH

BY

D. C. L.

"A thing of beauty is a joy for ever."

East view of the new Belfry, Ch. Ch., as seen from the Meadow.

SECOND THOUSAND.

Oxford:

JAMES PARKER AND CO.

1872.

FIGURE 46 Cover of a Lewis Carroll monograph

"beautiful, becoming, meet," and the German *frei,* meaning "free, unfettered, secure, safe." Therefore it is equivalent to "meatsafe," an object to which the belfry bears a perfect resemblance. Dodgson also speculates on why the design was chosen: Some say a chemistry student suggested it as a model of a crystal, but others affirm that a lecturer in mathematics found the design in the eleventh book of Euclid. The true story, says Dodgson, is that the belfry was designed by a wandering architect, now in a mental institution, who took his inspiration from a tea chest.

To get the best view of the belfry Dodgson recommends looking at it from one corner, so that one can see the edges of the cube converge in perspective on a vanishing point. This view gives rise to his happy thought: "Would that *it* were on the point of vanishing." Next, one should make a slow circuit around the quadrangle, "drinking in new visions of beauty at every step," and then walk slowly away until one experiences "the delicious sensation of relief" when the belfry is no longer visible.

The belfry's stunning design, Dodgson continues, has already inspired manufacturers. Two builders of bathing machines at Ramsgate are making their machines cubical, and there is now a bar of soap "cut in the same striking and symmetrical form." He has been told that Borwick's Baking Powder and Thorley's Food for Cattle are sold in no other shape, and he proposes that at the next Gaudy Night banquet each guest be given a "portable model of the new Belfry, tastefully executed in cheese." There is much more, including syllogisms, a dramatic skit and parodies of passages from famous poems. The complete monograph, along with Dodgson's drawing of the belfry, is reprinted in the Dover paperback *Diversions and Digressions of Lewis Carroll.* An introduction explains some of the monograph's inside jokes and topical allusions.

A remarkable polyhedron that would make a work of minimal sculpture far more interesting than a cube was discovered in 1977 by Lajos Szilassi, a Hungarian mathematician. It is a seven-faced toroidal polyhedron, that is, all its faces are polygons and it is topologically equivalent to a doughnut. It shares with the tetrahedron the extraordinary property that every pair of faces have an edge in common. Until Szilassi's computer program found the structure it was not known that it could exist.

In chapter 11 of my *Time Travel and Other Mathematical Bewilderments* (W. H. Freeman, 1988), I describe a 14-faced polyhedron that was discovered in the late 1940's by another Hungarian, Ákos Császár. The Császár and Szilassi polyhedrons are closely related. The Császár polyhedron is also a toroid, and it shares with the tetrahedron the property of having no diagonals. The Szilassi polyhedron is the topological dual of the Császár polyhedron: the two have the same number of

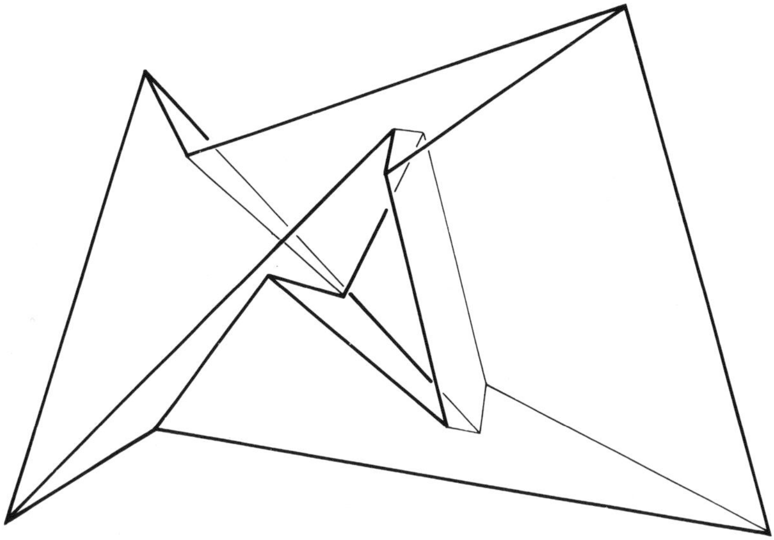

FIGURE 47 The Szilassi toroidal polyhedron

edges (21), but in the Szilassi polyhedron the 14 faces of the Császár polyhedron have been replaced by 14 vertexes and the seven vertexes of the Császár polyhedron have been replaced by seven faces. Figure 47 shows what the Szilassi polyhedron looks like. Note that the hole is unusually large and that there are three pairs of congruent faces. For readers who wish to make a model of the Szilassi polyhedron, E. N. Gilbert of Bell Laboratories has provided the patterns for all seven faces shown in Figure 48. Each pair of congruent faces can be cut as a single piece and folded along their common edge.

The most delightful aspect of the Szilassi polyhedron is that it shows how a seven-color map can be drawn on a torus, that is, a map that must be colored with seven colors so that no two adjacent regions are the same color. On the plane or on a solid topologically equivalent to a sphere the largest number of regions that can be mutually adjacent is four, a fact displayed by the four faces of a tetrahedron. On the torus the corresponding chromatic number is seven. Color each face of the Szilassi toroid a different color and then imagine the toroid inflated to the shape of a doughnut. The surface will be covered with a seven-region map requiring seven colors.

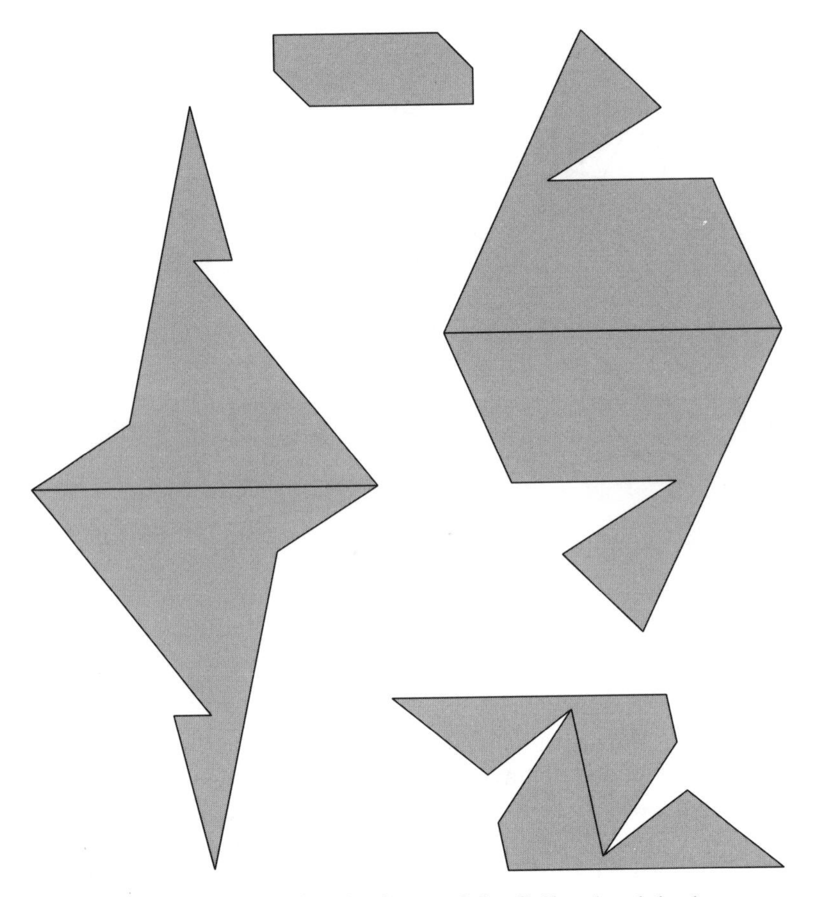

FIGURE 48 Patterns for the faces of the Szilassi polyhedron

Minimal sculpture is not, of course, limited to polyhedrons. Constantin Brancusi's *Bird in Space* is a well-known early example of free-form minimal sculpture. Many other sculptors have created simple structures with curved lines that have mathematically interesting properties. Eero Saarinen's mammoth Gateway Arch, which dominates the skyline of St. Louis, comes at once to mind. It has the form of an inverted catenary, the curve assumed by a chain when it is held at the ends and allowed to hang in a loop.

For the past forty years few artists have been more influenced by mathematical concepts than the Swiss artist Max Bill, as is shown by his hard-edge painting and minimal sculpture. Bill's fascination with topology is reflected in dozens of works featuring curved surfaces that, like the Möbius strip, are one-sided. Many of his constructions are strange but oddly pleasing dissections of a simple solid, such as a torus, a sphere or a cube, into two congruent parts. For example, in the work shown in Figure 49 Bill cut a black diorite torus in half, then balanced one part on the other.

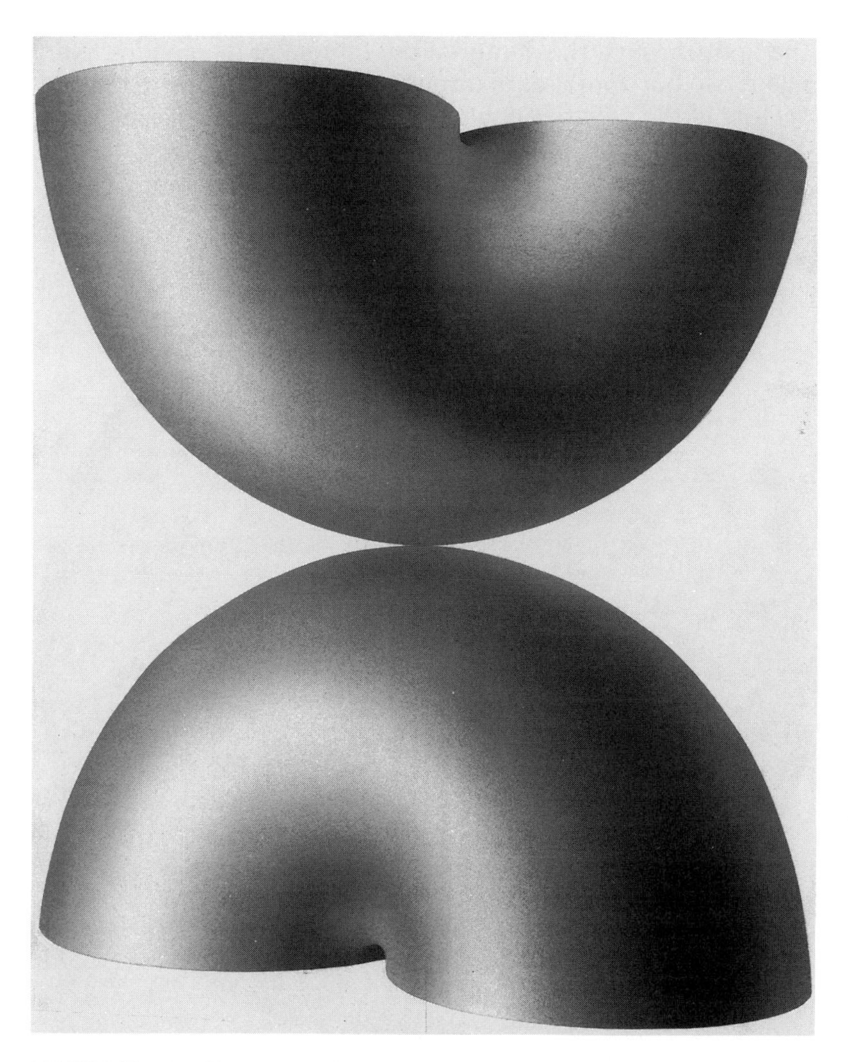

FIGURE 49 Drawing of Max Bill's *Construction from a Circular Ring*

Bill also executed a series of five works, each of which is based on a different way of cutting a sphere into two identical parts. Bill may not have been aware of it, but one of the works, the gray granite *Half Sphere around Two Axes,* is based on an old folk method of quickly slicing an apple into congruent halves. A half sphere cut by this method is shown in Figure 50. It is not as easy to make as it looks. Make a vertical cut halfway through the center of the top of a sphere. Turn the sphere over and make a second halfway cut, perpendicular to the first, through the center of the bottom of the sphere. Now make two horizontal cuts through diagonally opposite quarter sectors of the sphere's equatorial disk. The half sphere shown in the illustration is one of the two identical halves that re-

FIGURE 50 Half of a sphere

FIGURE 51 Minimal sculpture by Mitsumasa Anno

sult. Note how it suggests a three-dimensional version of the yin-yang bisection of a circle into asymmetric, congruent parts.

I close with a truly wondrous example of minimal sculpture designed by Mitsumasa Anno, a Japanese graphic artist. Figure 51 shows a work from his book, *Anno 1968–1977*. A huge marble version of this mysteriously truncated cube would make an appropriate monument for the sunny grounds of California's Stanford Research Institute. It would symbolize the management's faith in the military applications of their continuing research on the paranormal.

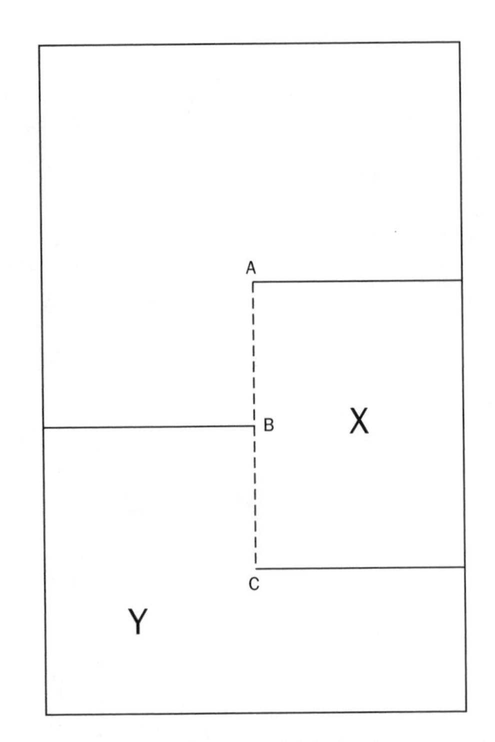

FIGURE 52 How to fold the hypercard

ANSWER

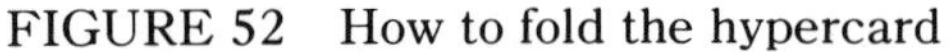

The "hypercard" is easily constructed as follows: Take a rectangular sheet of paper or cardboard and make three cuts along the black lines shown in Figure 52. Now fold flap *X* up 90 degrees along the broken line *ABC* and then turn the lower portion *Y* over by folding it back 180 degrees along the broken line *BC*.

ADDENDUM

The hypercard made a big hit with readers. For several weeks after the previous chapter appeared in *Scientific American* in 1978, you could see hypercards all over Manhattan, especially on art and advertising office desks.

Magicians who like mathematical curiosities had a field day with this seemingly impossible object. It occurred to those who enjoy practical jokes to make the hypercard from what are called double-face or double-back cards. Nonmagicians, unaware that cards are printed with backs on both sides, or faces on both sides, were twice as puzzled by such a hypercard, and of course unable to make one from an ordinary card.

A similar idea, which I first heard about from Tom Ransom, a Toronto amateur magician, was to carefully rule part of one side of a blank 3×5 file card with the familiar blue lines and one red line. Doing the same thing on part of the card's back enables you to produce a hypercard that is blank on its underside, with ruled lines on the top side and on one side of the mystery flap. It can be exhibited to guests, showing the blank underside before you tape it to the floor or a table.

Canadian magic buffs Howard Lyons, Mel Stover and Warren Stephens worked out the following way to present the hypercard as a magic trick. They had cards printed front and back as shown in Figure 53. The dotted line is not printed, but indicates how the card is pre-cut before doing the trick. When you display the card, your thumb conceals this cut. Using scissors, cut along the two black lines that have arrows pointing toward them, taking care not to reveal the cut under your thumb. Bend the flap back and forth, and show both sides of the card, while you patter about a man who wanted to cut a rectangle of plywood to produce a flap. After cutting the flap, he decided it left too large a hole. While you keep bending the flap back and forth, secretly give the card a twist to form the hypercard.

Put the hypercard on the table, with the flap folded flat and held down by your thumb to conceal one of the gaps. "Being

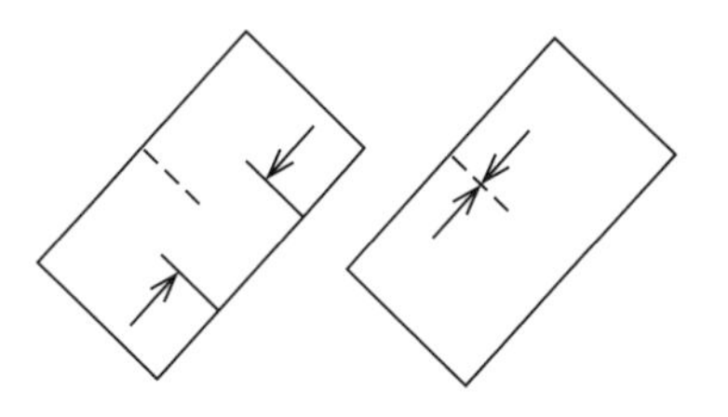

FIGURE 53 Front (left) and back of printed cards used in a hypercard magic trick

a magician," you continue, "the man had no difficulty reducing the size of the hole by half, as you see. In case you are wondering what happened to the other half of the hole, it's over here." Raise the flap to disclose the other gap. "What I can't understand is how such a flap could be cut from the two gaps and still stay in one piece!"

Allow the audience plenty of time to puzzle over the hypercard before you say "Beats me!" and tear up the card.

Karl Fulves suggested a similar presentation with a precut double-face or double-back card. As in the above trick, you keep the cut concealed under your thumb while you make the two cuts on the other side. The card is then secretly twisted (you can do this behind your back if you like) to produce the hypercard. Staple it to an ordinary playing card and hand it out for your victims to mull over.

Fulves also suggested the following variant of the Canadian trick. There are no arrows on the card. The cut made in advance is concealed by your thumb while you make the other two cuts. Fold the flap over the pre-cut to conceal it, and have someone initial the upper left corner as shown in Figure 54. You initial the other corner. Unknown to your viewers, you have previously initialed the opposite side of this corner. Move the flap back and forth, then secretly twist to make the hypercard. Both initials seem to be where they were before. Allow everyone time to study the impossible structure before you tear it up or staple it to an uncut card.

The hypercard has been used several times as a novelty give-away. Its first such use, as far as I know, was by the Office of Continuing Education, Ferris State College, Big Rapids, Michigan. Paul Merva, of that office, sent me a printed brochure of four pages formed by folding a sheet in half. The

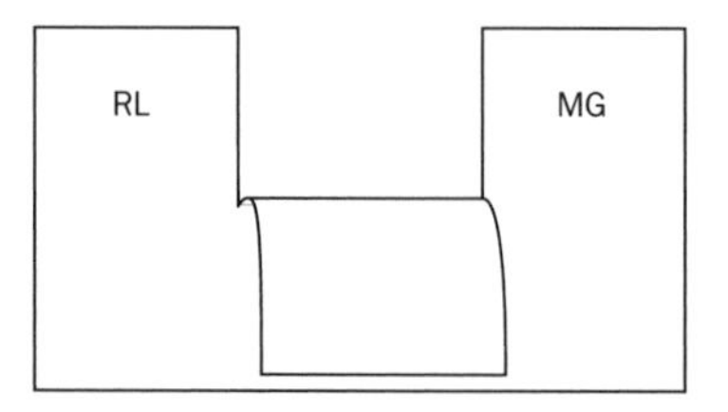

FIGURE 54 How hypercard, before twisting, is initialed in Karl Fulves's magic trick

brochure's second leaf was a hypercard. After the three cuts and twist had been made, the lower part of the card was so carefully glued back to the first leaf as to be indistinguishable from a fold. Of course the leaf had been printed so that the text read normally on both sides, with the flap's text providing the copy missing from the two sides.

Stover proposed using hypercards as placecards, each mounted with glue on transparent plastic. A message can be printed on the bottom of each card, to be read through the plastic, and the person's name written on the mystery flap.

Paul Merva and Alexis Gilliand each sent a model of the paper ring shown in Figure 55. It is actually a Möbius surface, and in the form depicted it seems even more mysterious than a hypercard. Equally perplexing is a band with two mystery flaps. They can be put in an untwisted band without having to cut it—simply make the three cuts on opposite sides of the band and give it two half-twists. A curious conversation piece can be made in this way from a seamless band obtained by cutting around a cereal box. The printing reads normally all the way around, and the two flaps, projecting outward from opposite sides, make it look like a truly impossible object. The same can be done with a cardboard napkin ring, or any ring cut from a paper or cardboard cylinder. Of course any

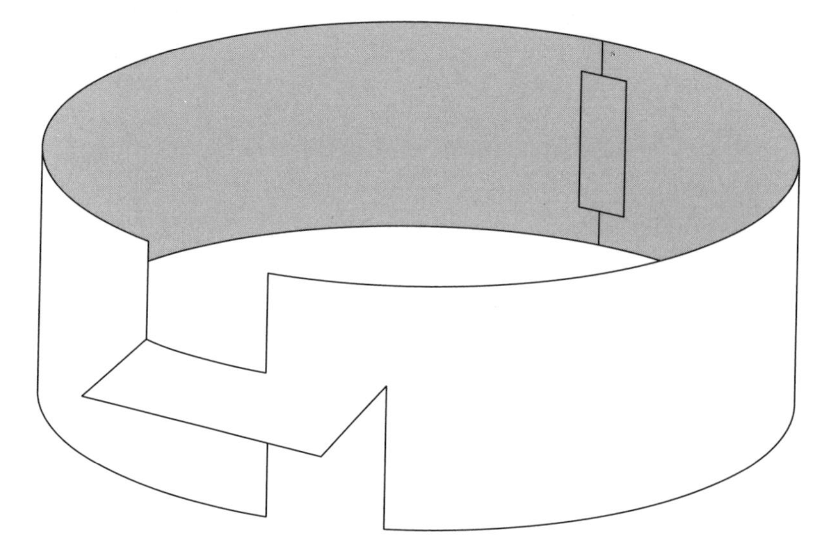

FIGURE 55 Mystery flap on a Möbius band

even number of flaps can be created in such a band by giving it an even number of half-twists, or any odd number of flaps by opening the band and giving it an odd number of half-twists before rejoining the ends. Gilliand also sent models of Klein bottles, with mystery flaps, that he constructed with paper.

Jack Botermans, in his book *Paper Capers,* explains how to construct three unusual variants of the hypercard, all with interior holes (see Figure 56). The places where cuts are made, and the paper rejoined, can be concealed in several ways: by carefully gluing and rubbing the glued line, by drawing decorative patterns on the cards, or by pasting something over the lines. Tricks Company, Ltd., a Tokyo magic supply firm, sells the hyperdisk shown in Figure 56. It was designed by Yasukazu Niishiro. Several five-pointed silver stars are pasted on the disk to conceal the two glued lines. Two of the stars con-

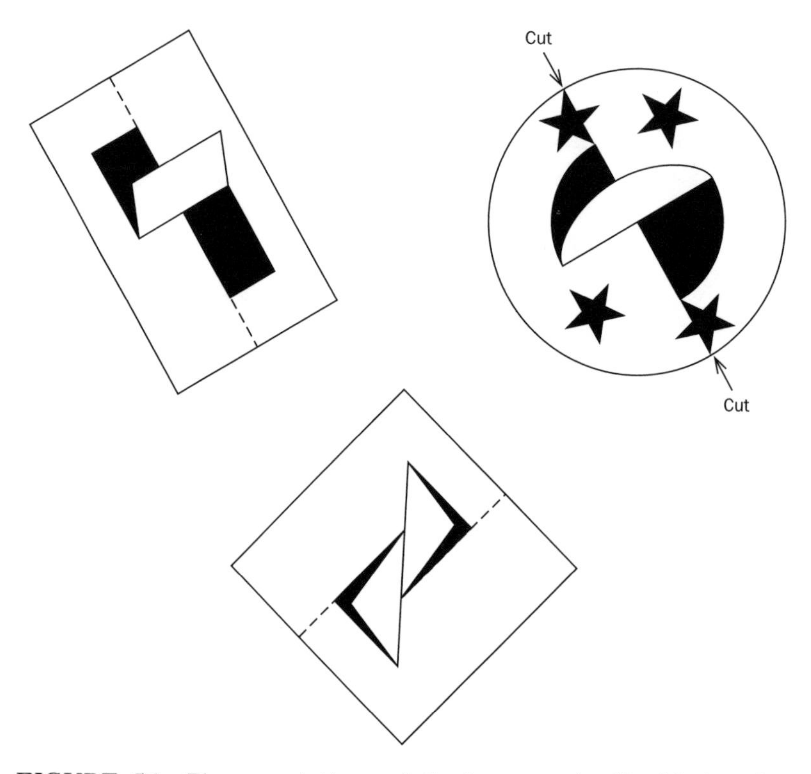

FIGURE 56 Three variations of the hypercard, all with interior holes (shown shaded). The cuts are shown dotted.

ceal the rejoined cuts as shown in the illustration. A transparent plastic cover, shaped like a pyramid, prevents one from examining the impossible structure.

The most elaborate presentation of magic tricks based on the hypercard is covered in Ben Harris's booklet, *The Hypercard Experience* (see the bibliography).

The method I described for bisecting an apple proved to be more interesting than I suspected. In France, it turns out, this curious way to slice an apple has long been a parlor stunt known as *la coupe du roi* (the cut of the king). Although each half of the apple is mirror asymmetric, the two halves are not mirror images of one another as one might suspect, but identical, like the congruent yin and yang of the Oriental symbol. Indeed, it is not possible to cut an apple (or a sphere, right circular cone, or right circular cylinder) into two asymmetric halves that are mirror images.

I found this out from a fascinating paper by four chemists at Princeton University entitled, "*La Coupe du Roi* and Its Relevance to Stereochemistry: Combination of Two Homochiral Molecules to Give an Achiral Product," by Frank Anet, Steve Miura, Jay Siegel, and Kurt Mislow. There had been no known example of a molecule formed by joining an asymmetric molecule to another molecule of the same sort, to produce a symmetric molecule analogous to the apple before it is cut. The four chemists succeeded in synthesizing just such a molecule.

I mentioned only Max Bill as a sculptor who relies on mathematical structures. Many other talented sculptors are producing mathematically based objects. Charles O. Perry, of Norwalk, Connecticut, is a sculptor whose work has been strongly influenced by topology. In England, Ronald Brown has been similarly influenced, especially by the topology of knots. Helaman Rolfe Pratt Ferguson, a professional mathematician at Brigham Young University since 1971, is rapidly gaining recognition as a serious sculptor.

Ferguson's work is of more technical interest to topologists than the works of any other sculptor known to me. His forms include a variety of toruses, Möbius surfaces, Klein bottles, cross-caps, knots and wild spheres such as the Alexander horned sphere, which he has modeled in several different ways (see Figure 57). His "Torus with Cross-Cap and Vector Field," a white marble work now at the headquarters of the American Mathematical Society, in Providence, Rhode

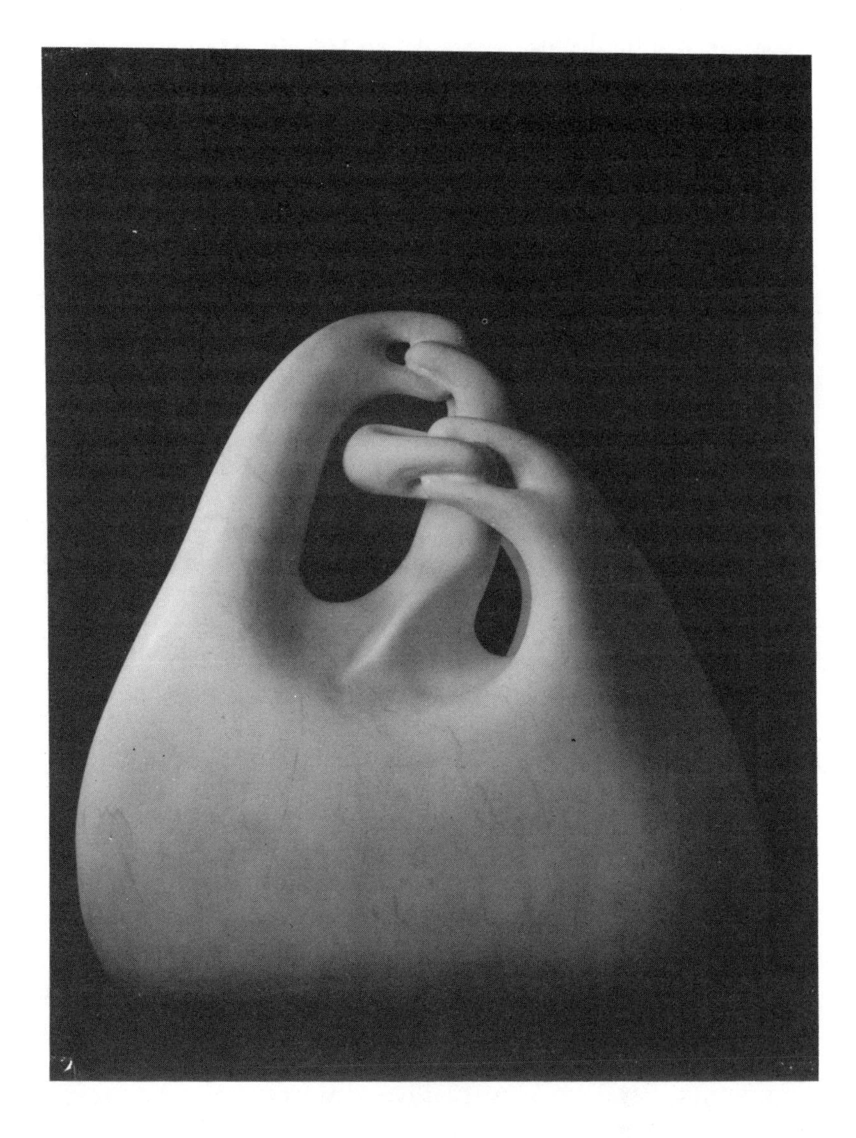

FIGURE 57 *Whaledream II,* Alexander's horned sphere, carved out of white carrara marble. Courtesy of Helaman Ferguson.

Island, is surely the first sculpture ever to show a cross-cap. Other Ferguson works are on university campuses, or owned by corporations and private collectors. See the bibliography for references on all three artists.

BIBLIOGRAPHY

"Packing a Box with Y-Pentacubes." C. J. Bouwkamp and D. A., Klarner, in *Journal of Recreational Mathematics,* 3, 1970, pages 10–26.

On hypercards

"Hypercard." Karl Fulves, in *The Chronicles,* Vol. 1, Nos. 1 and 12, 1978.

The Hypercard Experience. Ben Harris. Queensland, Australia: Ben Harris Magic, 1965.

The Ultracard Principle. Gordon Jepperson. Cayuga Falls, Ohio, 1985. A privately issued magic trick.

Paper Capers. Jack Botermans. Henry Holt, 1986. A translation from the Dutch.

On the apple cut

"La Coupe du Roi and Its Relevance to Stereochemistry." Frank A. L. Anet, Steve S. Miura, Jay Siegel, and Kurt Mislow, in the *Quarterly Journal of the American Chemical Society,* March 23, 1983.

"Merging Molecules of the One 'Hand.' " *Science News* (unsigned), 123, 1983, page 278.

On Max Bill

Max Bill. Introduction by James Woods, text by Lawrence Alloway. Buffalo Fine Arts Academy and the Albright-Knox Art Gallery, Buffalo, 1974.

"Max Bill: Superb Puritan." Robert Hughes, in *Time,* 104, 1974, pages 100–101.

On Isamu Noguchi

A Sculptor's World. Isamu Noguchi. Harper and Row, 1968.

"Isamu Noguchi, A Kind of Throwback." Harold C. Schonberg, in *The New York Times Magazine,* April 14, 1968.

Isamu Noguchi. Sam Hunter. Abbeyville Press, 1978.

"Isamu Noguchi's Elegant World of Space and Function." Benjamin Forgey, in *Smithsonian,* April, 1978, pages 46–55.

On Charles Perry

Charles O. Perry. Charles O. Perry. Privately published, Norwalk, Connecticut, 1987. A 64-page paperback with color photos of Perry's work since 1964.

On John Robinson

"The Sculpture of John Robinson: Reactions of a Mathematician." Ronald Brown. University College of North Wales, Mathematical preprint 90.23, July, 1990.

On Helaman Ferguson

"Mathematical Ideas Shape Sculptor's Work." Barry A. Cipra, in *SIAM News,* May, 1990, pages 24ff.

"Equations in Stone." Ivars Peterson, in *Science News,* 138, 1990, pages 150–154.

"Mathematics in Marble and Bronze: The Sculpture of Helaman Rolfe Pratt Ferguson." J. W. Cannon, in *The Mathematical Intelligencer,* 13, 1991, pages 30–39.

9

Minimal Sculpture II

I originally intended to add some remarks to the previous chapter's addendum about minimal sculpture, still bemusing museum directors, art critics and art historians; but my observations grew to such lengths that I decided to make a separate chapter out of them.

Minimal art, whether painting, sculpture or music, seems to me aptly named because it is minimal in more than one way:

1. It requires minimal time, effort, thought and talent to produce.

2. It can be constructed, in most cases, with a minimum of expense.

3. It has minimal aesthetic value.

Naturally it has *some* aesthetic value. But all around you, in tens of thousands of different forms, are objects with aesthetic value: trees, clouds, cats, people, buildings, cars, chairs, spoons, and so on. Almost anything has *some* aesthetic value. A pure expanse of a color has aesthetic value, but what is the point of coloring an entire canvas with a solid color (as minimalist painter Ad Reinhardt did), framing it, and hanging it in a gallery? A paperclip has some aesthetic value, but is it worth erecting as a statue in a public park? Does minimal sculpture have enough aesthetic value to justify the enormous prices cities and museums pay for it? Is it worth the opaque prose that streams endlessly from the pens of top art critics?

Picasso's Chicago statue is not as red as it once was, but Chicagoans have grown accustomed to its "face." In 1987 they celebrated its 20th birthday by hanging the city's Medal of Honor on the thing, and holding a street celebration complete with cake and ice cream. Nobody booed the way they did when the statue was first unveiled to a crowd of 50,000 on August 14, 1967. According to an Associated Press story (August 14, 1987), Chicagoans still haven't figured out whether Picasso intended it to represent a woman with a huge nose, an Afghan hound or a baboon. Art Critic John Canaday, guardedly praising the statue in the *New York Times* (August 27, 1967), found it "just plain ugly." If it proves to be an eyesore many years later, he concluded, it will be because its representative touches (eyes, nose, etc.) "become more and more distracting in a design that even now is marred by them." The Associated Press summed up the windy city's attitude with this verse:

Happy Birthday to you.
You're homely, it's true.
Just a rusty old sculpture,
But Chicago loves you.

I hope no one thinks I am fond of statues of war heroes on horses. They are indeed eyesores, but at least they have the merit of reminding us of our nation's history. Most minimal sculpture that I see in our big cities, and even in small cities, reminds me of nothing so much as the ugliness of city landscapes. With right angles and circles all around us, the mind longs for the chaotic randomness of clouds and trees—for sculptured forms that will not inflict more geometrical regularity on our tired retinas.

Take Carl Andre, for instance, whose pile of bricks I ridiculed in the previous chapter. He was born in Quincy, Massachusetts in 1935, and might have graduated from Kenyon University if the college hadn't expelled him after less than a semester. His longest lasting job, before he became world famous, was the four years he worked as a brakeman, later a freight conductor, for the Pennsylvania Railroad. He was fired after a mistake caused a freight-car accident. As I type I have beside me a handsome book titled *Carl Andre, Sculpture 1959–1977,* by David Bourdon. It is one of the funniest books of art criticism ever to come my way.

Andre's work, which has been exhibited in almost all the major museums of the United States, has one common denominator. It is made with modules—that is, identical, interchangeable units that are put together in various ways, but left unjoined, held together only by gravity. The modules are ready-made, "found" objects such as bricks, cement blocks, sheets of metal, styrofoam planks, timber, ceramic magnets, and so on. As Barbara Rose says in her preface to Bourdon's book, they return "to their original state" after their "death as works." Bourdon is surely accurate when he writes "Everybody's six year old child could probably recreate an Andre work and the replica would be indistinguishable from the original."

Early in his career Andre liked to stack his modules on top of each other the way children build towers with blocks, but then he shifted to his "flat period" in which the units are laid on the gallery floor or on outside ground. He wanted his sculpture to be "as level as water." Most of the time one could stand on his sculpture—for example, on his much admired 64 square steel plates arranged on the floor like a large chessboard (see Figure 58). Here is how Bourdon rhapsodizes over this work:

> The gestalt and materiality of the work are immediately apparent. There is no suggestion that the piece is any more than what it appears to be. The sculpture does not significantly change as we circumambulate it. Andre's decisions in composing the work are also self-evident: eight-inch modular particles are laid down in a square of eight rows of eight plates each. But the work holds the floor in a compelling way, functioning as a kind of zone within a larger space. Visually, the piece begins to come alive as we notice discrepancies and differences from one square to another; the various scratches, marks and variations in color and surface texture assume an almost auto-

FIGURE 58 Carl Andre's *64 Steel Square* (1967). Courtesy of Paula Cooper Gallery, New York. Private collection, Florida.

graphic quality. If we stand on the piece it tends to slip away in our peripheral vision, while the disconcerting way in which the tiles sway under our feet persuades us that the sculpture is not as stark and inflexible as we may have first imagined.

Another typical Andre work is *Secant,* a photograph of which graces the cover of Bourdon's book (see Figure 59). This consisted of 300 feet of brick-shaped fir timbers laid end to end like freight cars to loop across the grounds of the Nassau County Museum of Fine Arts, on Long Island. That this work, like so many of Andre's other "linear" works formed by putting modules end to end, could have been influenced by Andre's familiarity with freight cars is not lost on Bourdon:

> Writers on Andre are generally tempted to draw parallels between his experience with the Pennsylvania Railroad and his subsequent sculpture. It is certainly easy to understand how his taste for regimented, modular components might have been affirmed by the interchangeable freight cars and the evenly spaced railroad ties. The rails and ties could have helped persuade him that his sculpture should be horizontal, parallel to the earth, rather than standing, totem-like, upon it. The adaptability of the track to its terrain and the convergence of two or more lines in particular locations undoubtedly sharpened Andre's perception of "place." The artist acknowledges his railroad ex-

FIGURE 59 Carl Andre's *Secant* (1977). Courtesy of Paula Cooper Gallery, New York.

perience as a strong influence on his work. Whatever he learned in the rail yards, Andre emerged a much better sculptor.

Andre's linear works also resemble roads. Ordinary roads, he once said, are ideal examples of beautiful sculpture. There is something to be said for this. Roads can indeed be beautiful when they wind about through lovely scenery. When Dorothy and her friends followed the yellow brick road to the Emerald City, it must have been a great aesthetic experience. But a row of bricks stretched across a museum floor?

Not all of Andre's great works can be walked on. His *Spill* consisted of 800 little plastic tiles scattered over a gallery floor. His much more famous *Stone Field Sculpture* (see Figure 60) was constructed in 1977 on a grassy spot in Hartford, Connecticut, at the corner of Main and Gold Streets, adjacent to a historic church and graveyard. This "sculpture" consisted of 36 boulders carried by a construction crew from a quarry in Bristol, then lowered by cranes on the grass to form a triangular pattern of rows starting with one boulder, then two, then three, and ending with a row of eight. (Mathematicians know 36 as a "triangular number.") The National Endowment for the Arts and the Hartford Foundation for Public Giving paid Andre $87,000 for this.

"Are you putting us on?" someone asked Andre. (*New York Times,* September 5, 1977.)

"I may be putting myself on," Andre replied. "If I've deceived you, then I've deceived myself. It's possible."

"This is worse than Stegosaurus," said Mayor George Athanson, referring to a gigantic, bright-orange, abstract model of a dinosaur that Alexander Calder had inflicted on the city a few years earlier. "It's just a bunch of rocks."

FIGURE 60 Carl Andre's *Stone Field Sculpture* (1977). Courtesy of Paula Cooper Gallery, New York.

Andre's intent, he told the press, was to "bring together geological time and human time—it's important that people remember that difference." The sculpture was designed, he added, "to extend the serenity" of the graveyard "into the bustling city."

Here is a paragraph from a news account:

> However, New York welcomed Mr. Andre's exhibits at the Guggenheim Museum and at the Museum of Modern Art. The "Reef," a row of salmon pink styrofoam planks displayed in the Guggenheim driveway, and the "Water Body," nine narrow strips of metal arranged in a "constructivist abstraction" at the bottom of a pool, as well as other Andre works were cheered by New York critics for their beauty and logic. One [John Russell] wrote in admiration for one of his sculptures: It "just sits there and minds its own business."

The boulders are still sitting there in Hartford, I regret to say, minding their own business.

Although I am discussing Andre's sculpture in a book about mathematics, actually his understanding of mathematics is severely limited. After the previous chapter ran in *Scientific American,* I received an amusing postcard from Andre in which he said he had been reading *Scientific American* for thirty years, but found my column beyond his comprehension. "My work," he told me, "owes more to my Swedish brick-laying grandfather than to my mathematics, which never got beyond grade school."

Nevertheless, Andre has always found numbers intriguing, especially square numbers and primes. He quite consciously used prime numbers in many of his works, such as *Lever,* in which 137 bricks were stacked side by side on the floor of New York's Jewish Museum, in 1966, to form a long bar that resembled a lever (see Figure 61). Andre is also impressed by number sequences, especially 1,2,3,4, ... in which you can count the bricks when they are in a straight line, or the number of bricks in each row of a triangular formation. *The Way North, East, and South* (Figure 62) must have taken Andre at least three minutes to create. Four is a square number, though why "west" is not in the title is puzzling.

Bourdon's final paragraphs are typical of the pompous chatter of art critics:

> Andre deliberated long and hard at every stage of his renunciatory art, and the result is an *oeuvre* of exceptional strength

FIGURE 61 Carl Andre's *Lever* (1966). Courtesy of Paula Cooper Gallery, New York.

and vitality. Several of his sculptures are classics of the Minimalist mode and represent a drastic culmination of the reductive impulse—an impulse that did not originate full-blown with the Minimalists of the '60s, but which had been developed and

FIGURE 62 Carl Andre's *The Way North, East and South* (1975). Photo by Geoffrey Clements. Courtesy of Paula Cooper Gallery, New York. Private collection, San Francisco.

honed by numerous predecessors, ranging from Malevich and Mondrian to Rothko, Newman and Reinhardt. The further minimalization, or essentializing, by artists of the '60s was achieved mainly in terms of eliminating metaphysical references and emphasizing a more empirical kind of rationality.

Andre's originality and his contribution to the reductive aesthetic is already a historic, certifiable fact. Originality and quality are not synonymous, however, and at times even antithetical. The enduring quality of Andre's work will be determined later by others. For myself, I can only admire the stringent clarity of the sculpture, its commanding visual presence and stark, forthright physicality. The works strike me as concrete distillations of a unique, contemporary sensibility.

Bourdon's last paragraph is almost as vacuous as this description of one of Andre's brick constructions by Rosalind Krauss in her book *Passages in Modern Sculpture* (Viking, 1977):

> Instead the fire-bricks remain obdurately external, as objects of use rather than vehicles of expression. In this sense the ready-made elements can convey, on a purely abstract level, the idea of simple externality.

New York Times critic John Russell, who informed us that Andre's bricks like to sit there and mind their own business, had this to say about Andre's sculpture (February 20, 1976):

> In its general operation, it is in fact closer to the prayer mats of Islamic art than to sculpture as we have usually known it. The Tate piece is several inches thick, but a prototypical Andre may well be made up of thin plaques of metal that lie almost flush with the floor. Walking across one, we tread a field of light.
>
> The thralldom of a sculpture by Mr. Andre is owed to the clarity of his intention, to the frank and unambiguous way in which the materials are assembled and to the way in which a specifically American gift for plain statement has been applied to situations that in ordinary life are confused and contradictory. An Andre "is what it is," as a 17th-century philosopher said, "and not another thing."
>
> Sculpture for most museum visitors is something that clamors for our attention, takes up a lot of space and doesn't always deserve it and semaphores its messages from a long way away. An Andre is the antithesis of all this. Where other sculptors set up as dramatists, rhetoricians and demon persuaders, Mr. Andre comes on as a hermit who is concerned only to set his own surroundings in order.
>
> His work stands in this sense for order, resolution and an absolute simplicity. It steals upon us when we least expect it. It has what was defined not long ago as the prime characteristic of good new art: that it makes people wonder at first if it is art at all. Once we have come to like it, we ask ourselves how we could ever have gotten on without it.

Martin Ries, professor of art at Long Island University, made this reply in the *New York Times* (March 17, 1976) to a previous letter complaining about Andre's pile of bricks:

> As a long-time admirer of minimal art and especially Carl Andre's work (I gave him one of his first exhibitions in the early 1960's when I was assistant director of the Hudson River Museum) I would like to reply to Mr. Hodge's letter of March 1,

> in which he criticizes the Tate Gallery's purchase of Mr. Andre's sculpture.
>
> Mr. Hodge says he can admire the artistic merits of Constable and Van Gogh but the pile of bricks is "just a pile of bricks." His comparison (or anti-comparison) would be more accurate if he said he can appreciate the smears of colored linseed oil by Constable and Van Gogh but not Andre's structural arrangement of fired clay.
>
> A lot of people find profound meaning in this abstract balance between the spiritual and the material, which manifests harmony, proportion and pure order; I think Mr. Hodge some day will enjoy this aspect of Mr. Andre's work as much as he now enjoys the expressionistic quality of the paintings of Constable and Van Gogh.

Barbara Rose, wife of artist Frank Stella, wrote the preface to Bourdon's book about Andre. She calls Andre "the last Renaissance Man I ever knew" basing this on the fact that in addition to his sculpture Andre also writes shaped verse, "cacophonous atonal piano music," "unperformable operas," and "lyrics to mock pop tunes." He also has written novels, she said, that "became more and more abbreviated until they were but a paragraph in length, anticipating the contemporary attention span." I am reminded of the politician's rule that no one can be elected president unless he can put his central message on a bumper sticker, and of that classic minimal poem "To Fleas":

Adam
Had 'em.

Andre stoutly denies that his bricks are symbolic of anything. "A brick is a brick," he likes to say. When *Art Forum* planned to run a cover photograph of an Andre work that consisted of 29 bricks (a prime number) in a row, they decided it would be simpler to put down their own bricks and photograph them. Andre was furious when he found out about this. The magazine killed the cover after Andre threatened to sue for $500,000 because, he said, "I didn't lay the bricks." An unidentified artist told *New York* magazine (see the bibliography) that an argument with Andre ended when Andre put his face up close to the man and shouted "Brick, brick, brick. . . ." 128 times before the man lost count. Perhaps he repeated the word 131 times, a prime number.

In 1985 Andre was arrested in Manhattan for murdering his wife. He was charged with pushing Ana Mendieta out of the hip-high, sliding-glass window of their one-bedroom 34th floor apartment on Mercer Street, in Greenwich Village. Two indictments were dismissed because certain evidence was deemed inadmissable. On February 11, 1988, a third indictment was settled without jury. The State Supreme Court judge ruled that Andre was not guilty because his guilt could not be proved beyond a reasonable doubt. (See the *New York Times,* February 12, 1988.)

Ana, Andre's third wife, was a 36-year old Hispanic abstract artist, born in Cuba. She and Castro had been fellow law students, and Andre shared her left-wing political views. When the National Endowment for the Arts gave her a fellowship, she came to the United States to advance her art career.

In the early 1970s, when "body art" was fashionable, she became its passionate promoter. In one exhibit, she lay in a tomb covered with flowers that seemed to grow out of her body. In one of her "rape series" works she lay nude in the woods, her rear covered with blood. In another, she was tied to a table, nude from the waist down, with blood on the floor beneath her. In still another, she lay on a hotel roof, her body covered by a bloody sheet on which was placed a cow's heart. Ironically, after her fall her body was in fact covered with a bloodstained sheet.

At Puffy's, an artist hangout bar at Harris and Hudson Streets, a beer pitcher was labeled "Carl Andre Defense Fund." One customer contributed a brick.

In recent years many angry controversies have erupted in big cities around the world over monuments that the public consider ugly. And not just minimal sculpture, but also pop representations such as Claes Oldenburg's mammoth baseball bat in Chicago, the 4½-story clothespin in downtown Philadelphia that cost taxpayers $429,000, or the 16-feet high paperclip in Frankfurt, West Germany. The most publicized controversy was over Richard Serra's *Tilted Arc* (see Figure 63). Serra is probably our best known minimalist sculptor. His *Tilted Arc* was a 12-feet tall, 120-feet long, 78-ton curved, slightly tilted wall of rusting steel that slashed across the Federal Plaza in lower Manhattan.

The arc had been commissioned in 1979 by the art-in-architecture program of the federal government's General Service Administration. The agency gave Serra $175,000 in taxpayer money for this work. When it was installed in 1981, a great hue and cry arose from workers in the area. Not only

FIGURE 63 Richard Serra's *Tilted Arc* (1981). Kim Steele/Time magazine.

did they think it ugly, but it blocked a view, forced them to detour around it, and invited offensive graffiti.

"The work is a meaningless blob, an eyesore that ruins rare open spaces," wrote a *Time* reader (June 24, 1985). "Its only possible function is as a barrier against terrorists." Another *Time* reader, in the same issue, suggested that holes be cut in the wall for pedestrians to walk through, and that the work be retitled *Tilted Arc with Holes.*

Prominent people defended the arc: George Segal, Jacob Javits, Bess Meyerson, Joseph Papp, and Joan Mondale, wife of former vice president, Walter. Although the *New York Times* was critical of the arc on its editorial pages, its art critics defended it vigorously. Michael Benson, for example (May 19, 1985) called it a work of "great complexity and imagination:"

> What also makes "Tilted Arc" appropriate to its site is its content. The work has a great deal to do with the American Dream. The sculpture's unadorned surface insists upon its identity as steel. The gliding, soaring movement recalls ships, cars and, above all, trains. As with many enduring works of American art and literature, behind the sculpture's facade of overwhelming simplicity and physical immediacy lies a deep restlessness and irony.

In 1985 Serra was knighted in France by Francois Mitterand.

So loud was the public outcry over Serra's tilted wall that in 1989, after annoying New Yorkers for eight years, it was dismantled in the dead of night and carted off to be reassembled on a motor-vehicle compound in Brooklyn. Serra, furious over the removal of what he called a "site specific" work, lost a costly lawsuit to prevent this action. The U.S. government had to spend $50,000 more in taxpayer money to remove the wall. *The Wall Street Journal* headlined an editorial, "Good Riddance."

I hope no one supposes that only philistines, uneducated and art ignorant, found the wall ugly. I could fill pages with quotations from artists, art critics, and art historians who regard all minimal art as truly minimal in lowering aesthetic values almost to zero. I content myself with quoting only from two well known journalists and intellectuals. George Will, in his *Newsweek* column (I neglected to date my copy) blasted Serra for his arrogance, and minimal art in general for being "anti-intellectual" and "enveloped by ludicrous intellectualizing." John Simon called Serra a megalomaniac, with a "lust for celebrity," and a reputation that rested solely on publicity. One can get nothing but "uglification" from Serra, he wrote

> . . . , regardless of what the chorus of today's incestuous art experts may warble to the contrary. It is a wretched age that perceives Serra, and countless others equally vacuous and untalented, as having anything to do with any art other than that of self-promotion. Thank goodness for the *vox populi,* which, unacquainted with current art criticism, has the honesty and courage to pronounce ugliness offensive. If the artist has moral rights, let them be exercised where they do not clash with the moral and aesthetic rights of the rest of the population. The person who seeks out a work of art, however questionable, in a museum, gallery, or private collection has every right to do so. But innocent multitudes who have a horror thrust upon them in a public place near their homes or offices may rightfully refuse to put up with it.

In 1907, San Jose, California, paid $8,000 for a minimal statue made of several steel plates bolted together and painted red, blue, black and white. After it was in place at a downtown plaza, some construction workers, assuming it was junk they overlooked, hauled it off to a scrap heap. In 1989 the persons running an Arts Festival in Atlanta paid sculptor Frederick Nicholas $3,000 for his imitation Andre—a huge pile of empty banana boxes. Those attending the festival kicked the boxes all over the lot before a garbage truck was able to cart them away. That same year David Mach, a London minimalist, sold the Brooklyn Museum twenty tons of unsold periodicals that he and his assistants took a week to install in huge columns. The museum was miffed when Mach refused to remove the magazines after the show.

"Removing," said Mach, "is not a creative thing. I've never taken them down. That's part of the deal."

BIBLIOGRAPHY

Minimalism: Art of Circumstance. Kenneth Baker. Abbeyville Press, 1989.

On Carl Andre

Carl Andre: Sculpture 1959–1977. David Bourdon. New York: Jaap Rietman, 1979.

"A Death in Art: Did Carl Andre, the renowned minimalist sculptor, hurl his wife, a fellow artist, to her death?" By Joyce Walder, in *New York* magazine, December, 1985, pages 38–46.

On Richard Serra

"The Trials of Tilted Arc." Robert Hughes, in *Time,* June 3, 1985, page 78.

Richard Serra, Sculpture. Rosalind Krauss et al. Museum of Modern Art, 1986.

Richard Serra. Ernst-Gerhard Guse, et al. Rizzoli, 1988.

"Arc Without Covenant." John Simon, in *National Review,* May 5, 1989, pages 30–32.

"Richard Serra: Our Most Notorious Sculptor." Deborah Solomon, in the *New York Times Magazine,* October 8, 1989, pages 39ff.

10

Tangent Circles

When circles are tangent to one another, hundreds of beautiful problems arise, many of which have worked their way into the literature of recreational mathematics. Some historic examples have already been discussed in previous book collections of my *Scientific American* columns. In this chapter I shall take up a few more. There is no room for the proofs, but if interested readers will take the theorems as challenges to find proofs, they cannot help but strengthen their understanding of elementary plane geometry.

We begin with a famous figure believed to have been first studied by Archimedes and known as the arbelos, from the Greek for "shoemaker's knife," because it resembles the blade of a knife used by ancient cobblers. It is the shaded region in Figure 64, bounded by the semicircles with diameters *AB*,

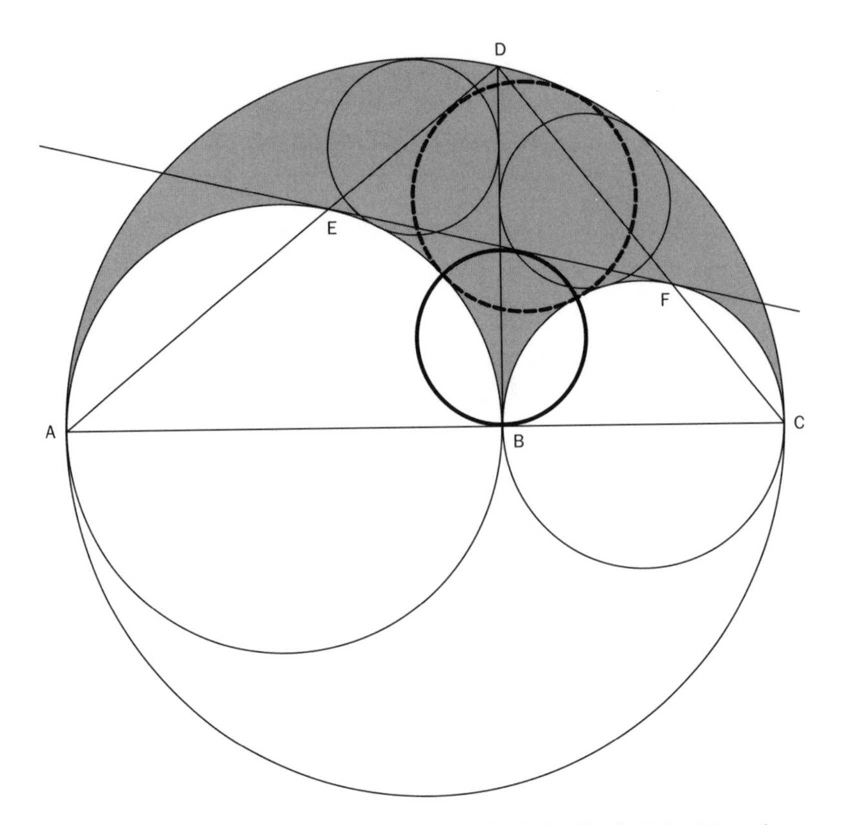

FIGURE 64 The arbelos, or "shoemaker's knife," of Archimedes

BC and *AC*. *B* can be any point on *AC*. Here are a few of the most amazing properties of the arbelos:

1. The length of semicircular arc *AC* is, as is easily shown, equal to the sum of the arcs *AB* and *BC*.

2. Draw *BD* perpendicular to *AC*. The area of the arbelos equals the area of a circle with diameter *BD*.

3. *BD* divides the arbelos into two parts. Circles inscribed in each part are identical, each circle having a diameter of $(AB \times BC)/AC$. The smallest circle circumscribing these twin circles also has an area equal to that of the arbelos.

4. Draw a line tangent to arcs *AB* and *BC*. Tangent points *E* and *F* lie on lines *AD* and *CD*.

5. *EF* and *BD* are equal and bisect each other. This ensures that the circle with diameter *BD* passes through points *E* and *F*.

6. In the early 1950's Leon Bankoff, a mathematician by avocation, made a curious discovery. (Bankoff is a Los Angeles dentist who describes himself as "Russian by extraction." He edits the problems section of *The Pi Mu Epsilon Journal,* and he is known for his skill in composing and solving problems.) Bankoff improved on the discovery, credited to Archimedes, of the twin circles shown on each side of BD in Figure 64 as follows: Draw a circle (shown in broken outline) tangent to the three largest circles, and then draw a smaller circle (shown in bold) that passes through B and the points where the broken circle touches arcs AB and BC. This circle also is identical with Archimedes' twin circles. Bankoff gives a proof in his article "Are the Twin Circles of Archimedes Really Twins?" No, they are not really twins, answers Bankoff; they are two circles in a set of triplets.

7. Construct inside the arbelos what is called a train of tangent circles. The broken circle in Figure 64 is the first in the train. It can be continued as far to the left as one wishes, in the manner shown in Figure 65. Label the circles C_1, C_2, C_3 and so on. The centers of all the circles in the train lie on an ellipse. The diameter of any circle C_n is $1/n$th the perpendicular distance from the center of that circle to the base line ABC. This remarkable result is in a fourth-century work by Pappus of Alexandria, who refers to it as an ancient theorem.

The proof of Pappus' theorem is simple if one uses inversion geometry, inverting the entire figure with A as the center of inversion. This converts semicircles AB and AC

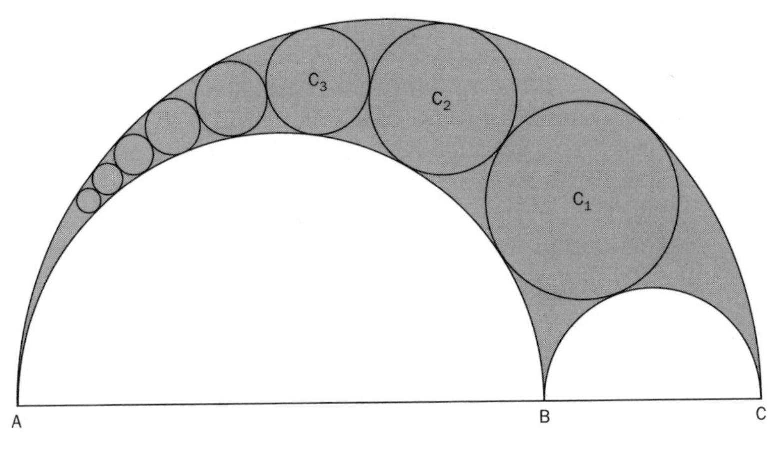

FIGURE 65 The arbelos train

to parallel lines, and the train becomes a set of equal circles bounded by the two lines. You will find a good explanation of how this is done in Rodney T. Hood's article "A Chain of Circles". Pappus did not know inversion methods (they were not developed until the 19th century) and so his proof is more cumbersome.

8. If AB equals 2 and AC equals 3, then the train of circles has many more surprises. The diameters of all circles in the train are rational fractions equal to $6/(n^2+6)$. Thus C_1 equals 6/7, C_2 equals 3/5 and so on. As reader Norman Pos pointed out in a letter, the center of C_2 lies on the diameter of the outside circle that is perpendicular to AC. Moreover, the centers of C_2 and C_3 are on a line parallel to AC, and that is also true of the centers of C_1 and C_6. The latter result is a special case of a more general theorem. If AB and AC are integral and AC equals $AB+1$, the centers of every pair of circles whose subscripts have a product equal to $AB \times AC$ lie on a line parallel to AC. Hence if AB equals 3 and AC equals 4, circle pairs with subscripts 1 and 12, 2 and 6, and 3 and 4 all have centers on a line parallel with AC. (For a proof see M. G. Gaba's 1940 paper, cited in the bibliography.)

9. If B divides AC in the golden ratio, many other striking properties result. These are discussed by Bankoff in his article "The Golden Arbelos." A note to publishers: Bankoff has an unpublished manuscript of 10 chapters on the arbelos, written in collaboraiton with the French mathematician Victor Thébault.

Closely related to the arbelos is a surprising theorem discovered by Jakob Steiner, a 19th-century Swiss mathematician, and depicted in Figure 66. A small circle is drawn anywhere inside a larger one, and in the region between the circles a train is inscribed. In most cases the train will not exactly close to form a ring of tangent circles, that is, the end circles will overlap. In some cases, however, the train will form a perfect ring like the one shown with solid lines in the illustration. When this happens, the train is called a Steiner chain. What Steiner discovered was that if the two initial circles allow one Steiner chain, they allow an infinite number of chains. Put another way, no matter where you draw the first circle of the chain, if you add the other circles, the chain will always close exactly. An arbitrary second chain is shown in the illustration with broken lines.

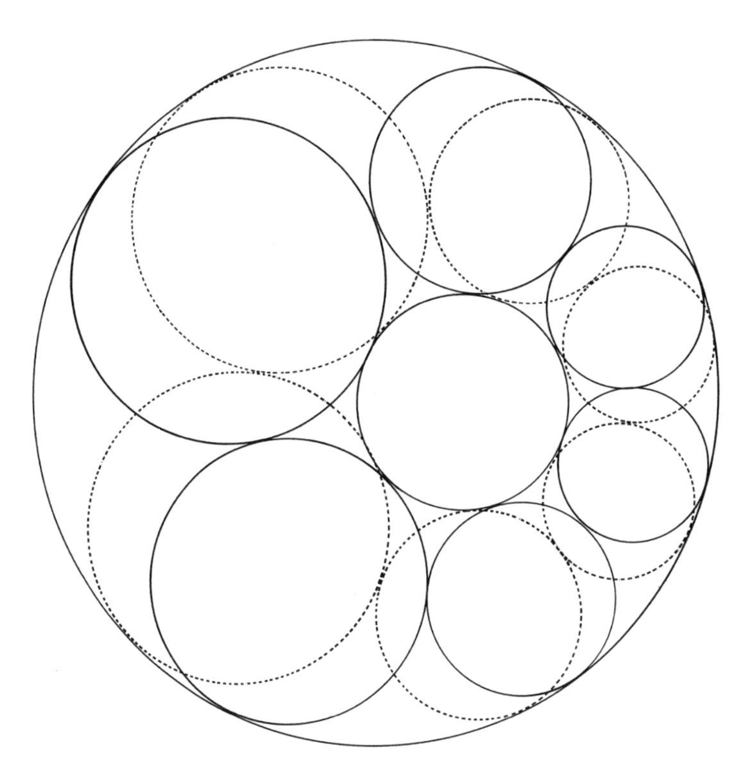

FIGURE 66 Jakob Steiner's chain

As before, the easiest proof is by inversion geometry. One performs an inversion that transforms the two initial circles into concentric circles. The Steiner chain then becomes a chain of identical circles that fill the region between the concentric circles. Details are given in the book by J. H. Cadwell, listed in the bibliography.

Solomon W. Golomb, whose contributions to recreational mathematics are widely known, was on a trip through Europe and found himself carrying a variety of coins of different sizes. The following thought occurred to him. Suppose n coins of varying size form a closed chain that exactly surrounds a central coin, as is shown in Figure 67. If the order of coins in the "wreath" is permuted, will the coins still form a perfect wreath?

Most people guess yes, and there is even a "proof." Draw lines from the center of the interior coin that go between each pair of adjacent coins in the wreath, as is shown in the illustration. The sum of all these central angles must be 360 de-

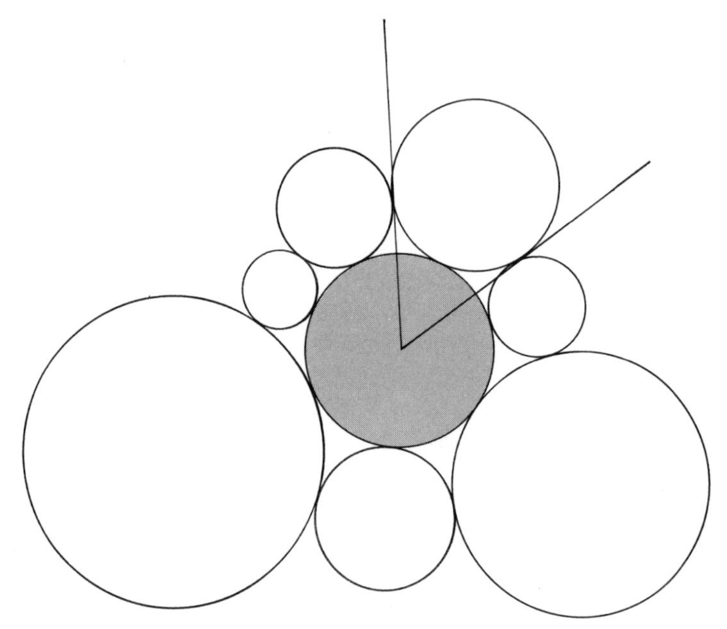

FIGURE 67 A false proof of Solomon W. Golomb's coin problem

grees, and this fact seems to be independent of the way the coins are arranged.

The proof is fallacious because, as Golomb points out, it assumes that the radiating lines must be tangent to each pair of coins they pass between. That is not always the case, however, and when it is not, the order of coins in the wreath can make a difference. Of course, if there are three coins in the wreath, permutations will have no effect because they merely give rise to rotations or reflections of the original figure. When there are four or more coins, it is easy, Golomb discovered, to find examples where the wreath closes in certain permutations but not in others.

Differences between permutations are slight unless there are large discrepancies in coin sizes. Therefore if you arrange a half-dollar, a quarter, a nickel, a dime and a penny around a central circle, you will find that any permutation of the five coins seems to fit exactly. Nevertheless, the differences are there. I leave it to readers to find ways of proving this statement correct. You might also like to tackle another one of Golomb's discoveries. Given n coins, no two of which are alike, what is the largest number of different interior circles they can exactly surround by permuting their order? The answer is $(n-1)!/2$. (The exclamation mark is a factorial sign.) Hence for four coins there are three permutations, for five there are 12, for six there are 60 and so on.

Golomb poses an interesting unsolved question. Given n coins of different sizes, what procedures will minimize and what procedures will maximize the size of the circle they can exactly surround? Golomb has some conjectures for both algorithms.

A completely different kind of problem about touching circles, not well known, involves the packing of n identical circles, without overlap, into a specified boundary of the smallest area. This problem has practical aspects, because cylinders such as cans and bottles are often packed in containers with circular, square or other cross sections. What is the smallest area of the cross section that will make it possible to pack n cylinders? To formulate the problem another way, given the area of a region and n identical circles, what is the largest diameter of the circles that allows packing them into the region without overlap?

No general solution is known, even when the boundary of the region is as simple as a circle, a square or an equilateral triangle, and in each case optimal packings have been established only for very low values of n. When the boundary is a circle, proofs are known only for $n=1$ through $n=10$; they were first given in a 1969 paper by Udo Pirl. The cases $n=2$ through $n=10$ (taken from "Packing Cylinders into Cylindrical Containers," by Sidney Kravitz, are shown in Figure 68. The minimum diameter of the outside circle is given below each figure, assuming that the small circles are of unit diameter.

Kravitz supplies the best solutions he could find empirically for $n=11$ through $n=19$. The case $n=12$ is of special interest. One would think that the close packing shown at the left in Figure 69 would be the densest, but Kravitz found the slightly better pattern shown at the right. Michael Goldberg, in "Packing of 14, 16, 17 and 20 Circles in a Circle," gives better packings for the four cases cited in his title. His packing for 17 circles was in turn improved by George E. Reis (see the bibliography) who also gives conjectured solutions for $n=21$ through $n=25$.

An inferior packing for $n=12$ was the secret of "The Packer's Secret," a popular puzzle sold in France late in the 19th century. The puzzle consisted of a circular box containing 12 checkers. The task was to pack them into the box in a

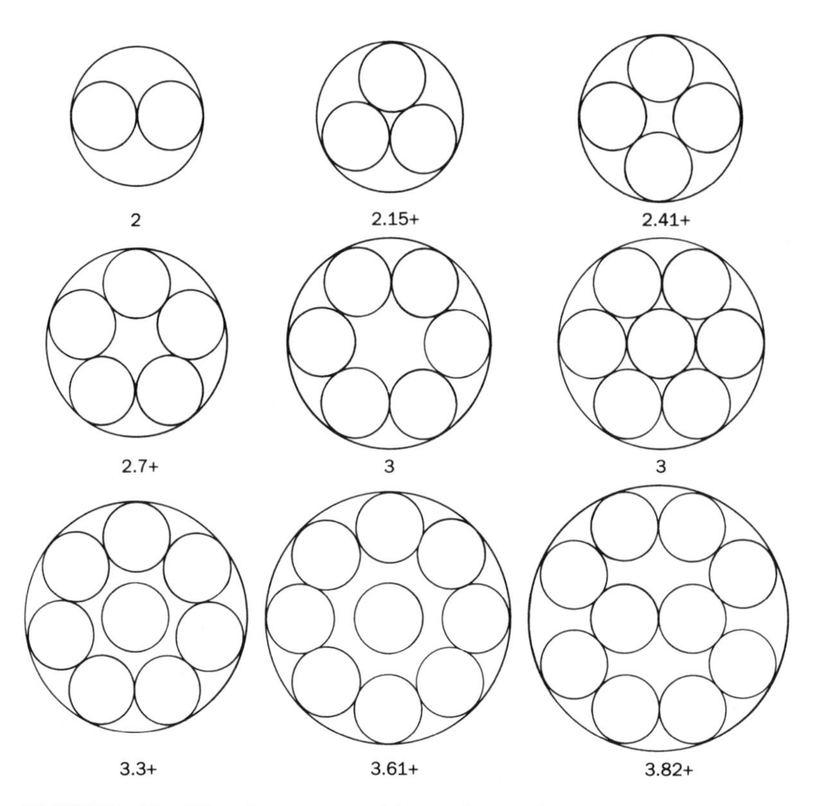

FIGURE 68 The densest packings of two through 10 unit circles into circles. The numbers below are the minimum diameters of the outside circle assuming unit circles inside.

stable, rigid way, so that if the box were turned upside down without the lid, the checkers would not fall out. Shown in Figure 70 is a circle just the right size for working on the Packer's Secret with 12 U.S. pennies. Can you fit 12 pennies into this circle to form a rigid pattern in which no coin is movable?

Searches have also been made for the densest packings of n identical circles into squares. It has been shown that as n increases, the density approaches .9069+. That is the limit obtained by the familiar close packing of circles with their centers on a regular lattice of equilateral triangles. Proofs of the best packings are known, however, only for $n=1$ through $n=9$. Once again there is no known formula or algorithm that yields the densest packing. Figure 71, reproduced from Gold-

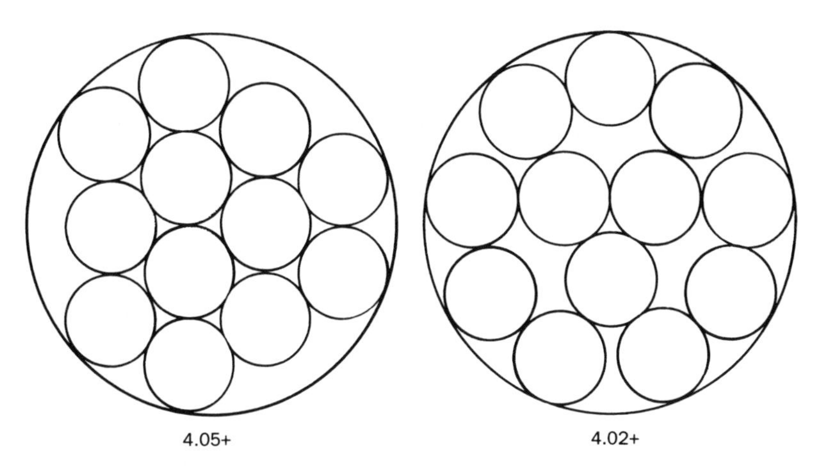

FIGURE 69 Inferior packing of 12 circles (left) and conjectured best packing (right)

FIGURE 70 Circle for working on "The Packer's Secret" with 12 U.S. pennies

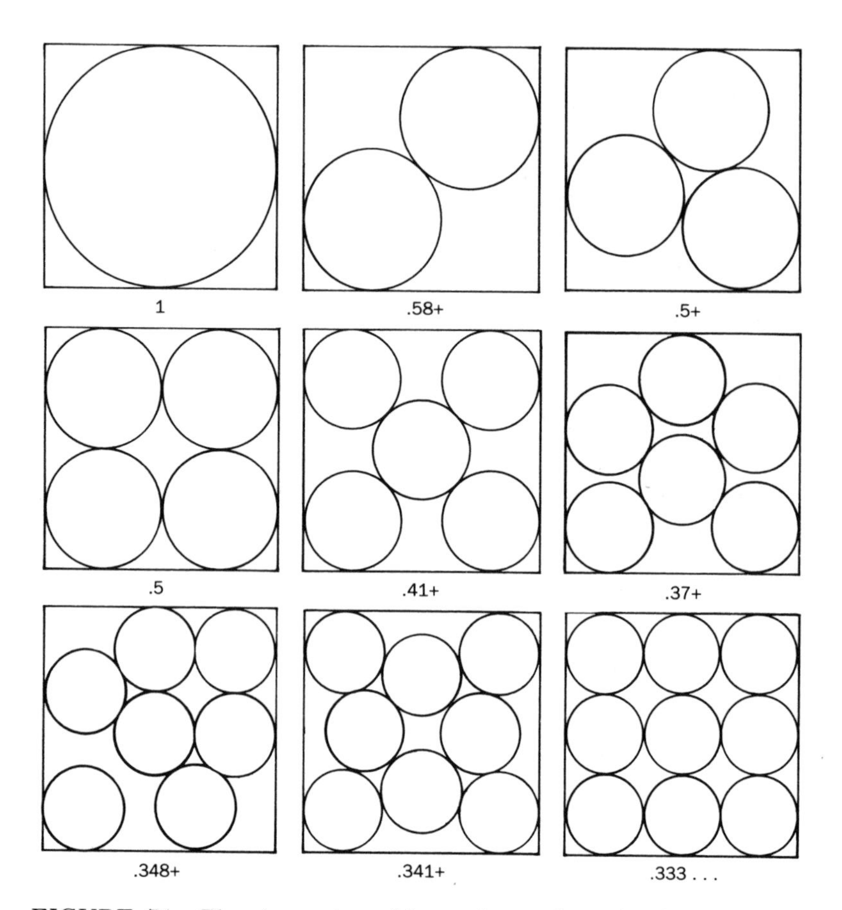

FIGURE 71 The densest packings of one through nine identical circles into squares. The numbers below are the diameters of the circles.

berg's "The Packing of Equal Circles in a Square," shows solutions for $n=1$ through $n=9$. In the illustration, instead of displaying unit circles inside the smallest square, Goldberg shows the largest circles that fit into a unit square. His paper also gives his best conjectures for $n=10$ through $n=27$, and for selected higher values.

Goldberg proves that in each best solution there must be a structure of touching circles that connects all four sides of the square, and that within this structure each circle must make contact with at least three other circles or a side of the square. There may, however, be circles that are not part of this structure, as in the case of $n=7$.

Below each square in the illustration is the diameter of the largest possible circle, assuming that the square's side is 1. In two cases, $n=6$ and $n=7$, proofs are not easy and have not been published. The case of $n=6$ was first solved by Ronald L. Graham of Bell Laboratories. The case of $n=7$ was announced by Jonathan Schaer in 1965. In a 1971 note in *Mathematics Magazine,* Schaer improved on Goldberg's conjecture of $n=10$. When the problem is extended to packing spheres into spheres or into cubes, it becomes enormously more intractable. Schaer has found solutions up to nine spheres, and Goldberg has made conjectures from nine to 27, and a few higher numbers. (Their two respective papers are cited in the bibliography.)

The packing of n equal circles into equilateral triangles also presents difficult questions. Little has been established except that when n is a triangular number (in the sequence 1, 3, 6, 10, 15 . . .), the densest packing is achieved by close packing in rows of 1, 2, 3, 4, 5 . . . circles. If the number of circles is a triangular number decreased by two or more, the remaining circles can always be shifted to fit into an equilateral triangle of smaller size. Hence if you remove two pool balls from inside the wood triangle used for closely packing 15 balls at the start of a game, the remaining 13 balls can be rattled around so that none of them touches the boundary.

Suppose just one ball is removed. Can the remaining 14 balls be moved about until none touches the boundary? It seems unlikely, but no one knows for certain. Donald J. Newman has conjectured that in all cases where the number of balls (or circles) is one less than a triangular number there is no way to rearrange the balls to make possible a smaller enclosing triangle. The conjecture does not apply to the triangular number 1, and it is clearly true for 3. It seems to be true for the next number, 6, but I know of no formal proof that five balls cannot be squeezed into a smaller triangle than the one that holds six.

We can ask similar questions about the densest packing of equal circles into any defined region, including regions with holes. If the region has no holes and is bounded by a convex closed curve, the best results obtained so far is by J. H. Folkman and Graham in "A Packing Inequality for Compact Convex Subsets of the Plane." Given the area and perimeter of a region, the authors establish an upper bound for the maximum number of unit circles that can be packed into it.

From hundreds of other theorems about touching circles, I have space for just one more: an elegant result published in 1968 by the Canadian geometer H. S. M. Coxeter. It is shown in Figure 72. An infinite sequence of circles is constructed so that every four consecutive circles are mutually tangent. It turns out that this sequence is unique when the ratio between consecutive radii is the one specified. The radius of each circle is obtained by multiplying the radius of the next-smallest circle by the sum of the golden ratio and its square root, a number that is slightly more than 2.89. The contact points of the circles lie on an equiangular spiral shown by the broken curve.

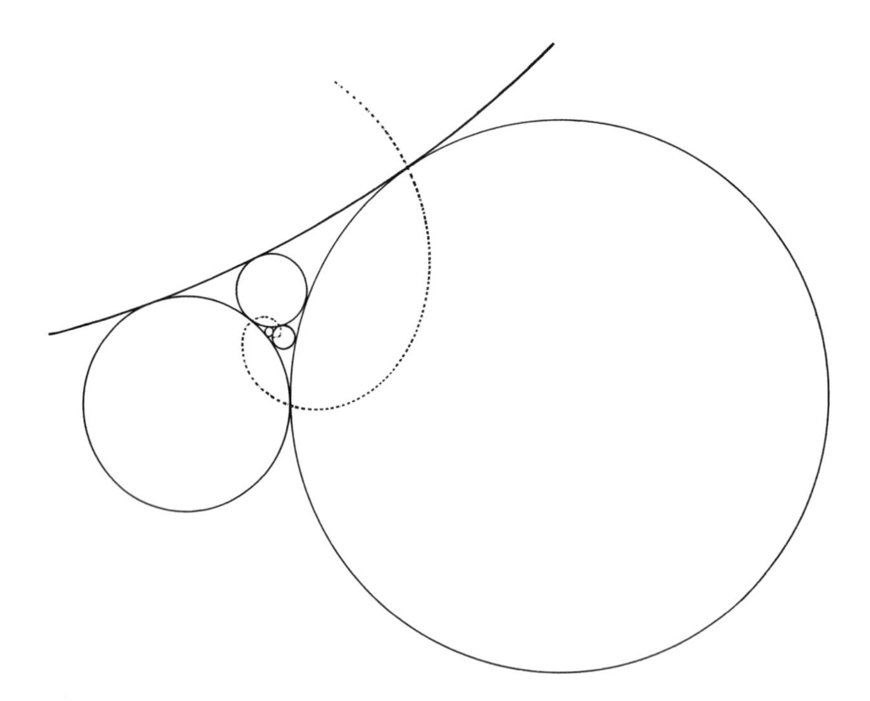

FIGURE 72 H. S. M. Coxeter's golden sequence of tangent circles

Answer

The solution to "The Packer's Secret" is shown in Figure 73. In doing the actual puzzle, it is expedient to start with one penny in the center; after 11 more have been placed around it, the center penny can be moved out to the rim.

Addendum

I should have explained that the problem of finding the densest packing of n circles inside a given figure is equivalent to

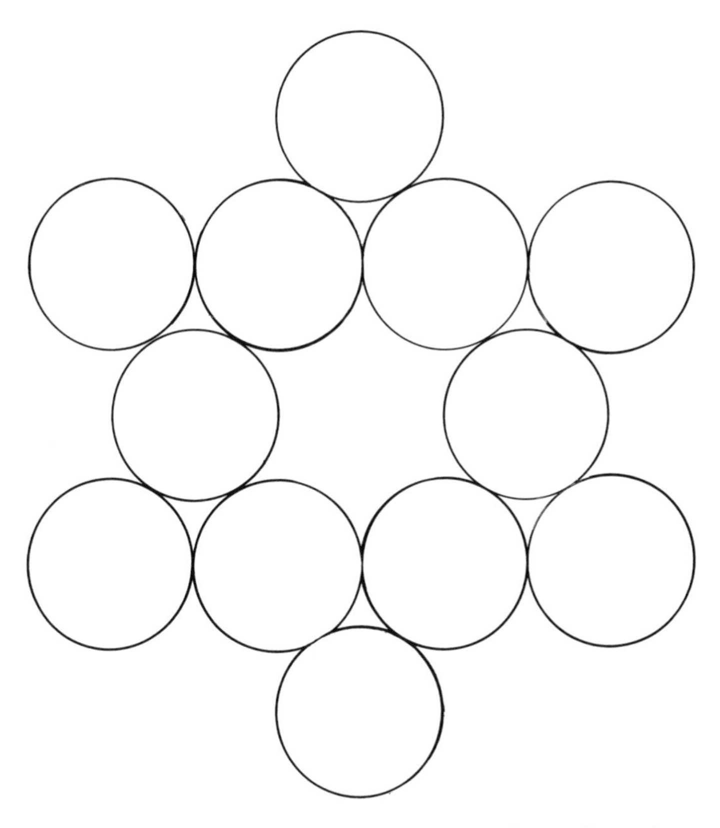

FIGURE 73 Solution to "The Packer's Secret" puzzle

finding the maximum number of points that can be placed inside the figure so that the smallest distance between a pair of points is as large as possible. Many papers on problems of this sort approach it in this form.

When the chapter was first published in 1978, I left open the problem of the densest packing of 10 circles inside a square. This has since had an interesting history. In 1970 the late Michael Goldberg found a stable symmetric pattern in which the radius of each circle is 5/34 or .1470+. This was slightly improved in 1971 by Schaer who found a symmetric but unstable pattern that raised the radius to .1477+. R. Milano, in 1987, made still further progress with a stable symmetric pattern which improved the radius to .1479+. He was able to prove that no better *symmetric* pattern is possible. In 1989 Guy Valette found an unstable asymmetric pattern which

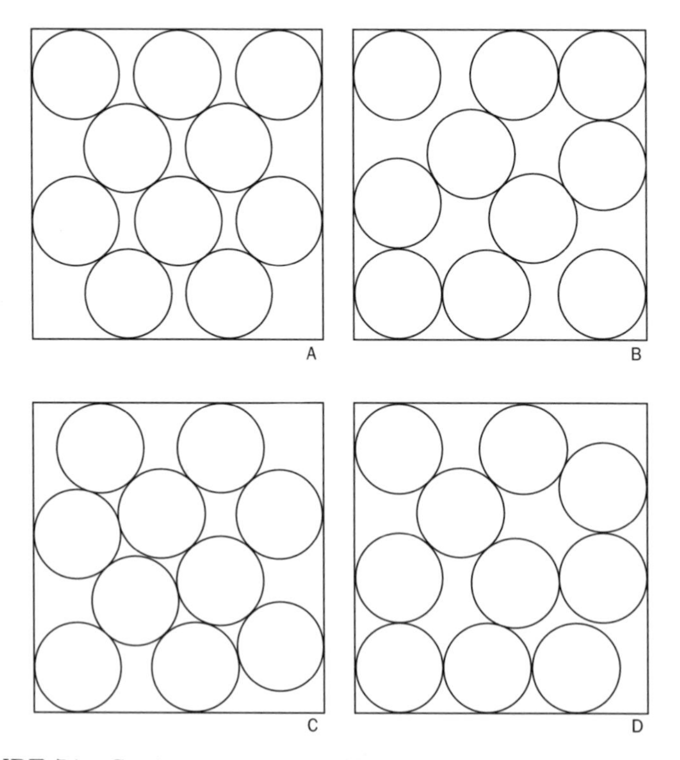

FIGURE 74 Conjectured solutions for ten circles by Goldberg (A), Schaer (B), Milano (C) and Valette (D).

lifted the radius to .1481821+. The four patterns are shown in Figure 74.

Valette conjectured that his solution was the best possible. This was shattered in 1990 by two French mathematicians, Michel Mollard and Charles Payan, who raised the radius to .148204+. Following the tradition of previous workers on the problem, they conjectured that no better solution could be found. This remains to be proved. Their packing is shown in Figure 75. Their paper also gives improved solutions for 11, 13, and 14 circles. Relevant papers are listed in the bibliography.

Goldberg called my attention to the fact that plastic covers for standard six-and-a-half ounce cans of nuts are just the right size for working on the problem of tight packing 12 pennies in a circle.

A famous circle packing problem that I did not have space to include is known as Malfatti's problem, after the Italian mathematician Gianfrancesco Malfatti who posed it in 1803 but failed to solve it. What is the maximum total area of three circles that can be packed without overlap inside a given right triangle of any shape? Malfatti's solution: the three circles that are tangent to each other, and each also tangent to two sides of the triangle. Such circles came to be called Malfatti circles. Many papers were published on how to construct such circles and calculate their sizes.

Not until 1929 was it discovered that Malfatti circles are not always the solution. In an equilateral triangle, for example, an inscribed large circle with two smaller ones be-

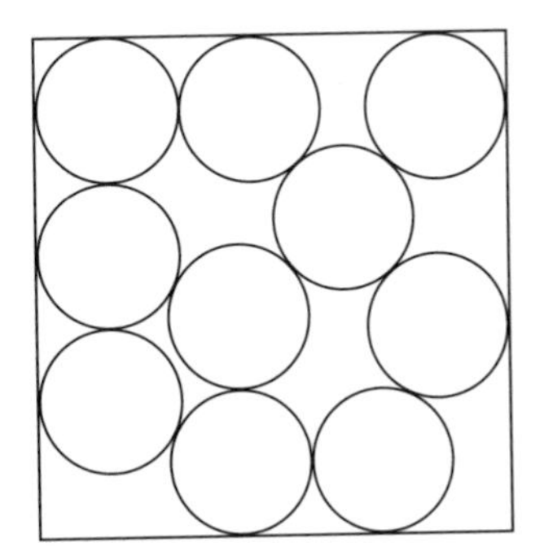

FIGURE 75 Improved 10-circle pattern by two French mathematicians. Is it the best?

tween the big circle and two corners have a combined area greater than the Malfatti circles. In a very tall isosceles triangle, three circles in a column will defeat Malfatti circles.

The final surprise came in 1967 when Michael Goldberg proved that Malfatti circles are *never* the solution! The following construction always beats Malfatti. Inscribe a circle in the given right triangle. Next, inscribe a circle tangent to the first circle and inside the smallest angle of the triangle. Finally, inscribe a third circle either in the same angle or in the next larger angle, whichever allows the larger circle. (In some cases, the two circles have the same combined area.)

Norman Pos wrote to tell me about another unexpected property of the arbelos. In Figure 65 when $AB=2$ and $AC=3$, the ellipse, which is the locus of the centers of the chain of circles, has its two foci exactly at the centers of the circles with diameters AB and AC.

Garry Ford discovered that six dimes and a quarter seem to exactly surround a quarter. What are the relative diameters of the circles? Ford discussed the problem in a *Technology Review* piece (76, 1974, pages 57–58).

Solomon Golomb's results on his problem of circles surrounding circles are given in his paper, "Wreaths of Tangent Circles," cited in the bibliography.

BIBLIOGRAPHY

Circles. Dan Pedoe. Pergamon, 1957.

"Loxodromic Sequences of Tangent Spheres." H. S. M. Coxeter, in *Aequationes Mathematicae,* 1, 1968, pages 104–121.

"A Packing Inequality for Convex Subsets of the Plane." J. H. Folkman and Ronald L. Graham, in *Canadian Mathematical Bulletin,* 12, 1969, pages 745–752.

"Wreaths of Tangent Circles." Solomon W. Golomb, in *The Mathematical Gardner,* David Klarner (ed.). Wadsworth, 1981.

"Penny-Packing and Two-Dimensional Codes. R. L. Graham and N. J. A. Sloane, in *Discrete and Computational Geometry,* 5, 1990, pages 1–11.

On the arbelos

"On the Circles of Pappus." Victor Thébault, in *American Mathematical Monthly,* 47, 1940, pages 640–641.

"On a Generalization of the Arbelos." M. G. Gaba, in *American Mathematical Monthly,* 47, 1940, pages 19–24.

"The Golden Arbelos." Leon Bankoff, in *Scripta Mathematica,* 21, 1955, pages 70–76.

"A Chain of Circles." Rodney T. Hood, in *Mathematics Teacher,* 54, 1961, pages 134–137.

Topics in Recreational Mathematics. J. H. Cadwell. Cambridge University Press, 1966.

"Are the Twin Circles of Archimedes Really Twins?" Leon Bankoff, in *Mathematics Magazine,* 47, 1974, pages 214–218.

On Malfatti's Problem

"On the Solution of Malfatti's Problem for a Triangle." H. Lob and H. W. Richmond, in *Proceedings of the London Mathematical Society,* 2, 1930, pages 287–304.

A Survey of Geometry, Vol. 2. Howard Eves. Allyn and Bacon, 1965, pages 245–247.

"On the Original Malfatti Problem." Michael Goldberg, in *Mathematics Magazine,* 40, 1967, pages 241–247.

"The Converse Malfatti Problem." Michael Goldberg, in *Mathematics Magazine,* 41, 1968, pages 262–266. On the task of finding the triangle of least area that can enclose three nonoverlapping circles of given radii.

On dense packing of circles in circles

"How Many Wires Can Be Packed Into a Circular Conduit?" Jacques Dutka, in *Machinery,* October, 1956, pages 245–247.

"Packing Cylinders Into Cylindrical Containers" Sidney Kravitz, in *Mathematics Magazine,* 40, 1967, pages 65–70.

"Packing of 14, 16, 17 and 20 Circles In a Circle." Michael Goldberg, in *Mathematics Magazine,* 44, 1971, pages 134–139.

"Dense Packing of Equal Circles Within a Circle." George E. Reis, in *Mathematics Magazine,* 48, 1975, pages 33–37.

On dense packing of circles in squares

"On a Geometric Extremum Problem." Jonathan Schaer and A. Meir, in *Canadian Mathematical Bulletin,* 8, 1965, pages 21–27.

"The Densest Packing of Nine Circles In a Square." Jonathan Schaer, in *Canadian Mathematical Bulletin,* 8, 1965, pages 273–277.

"The Packing of Equal Circles In a Square." Michael Goldberg, in *Mathematics Magazine,* 43, 1970, pages 24–30.

"Separating Points in a Square." Benjamin L. Schwartz, in *Journal of Recreational Mathematics,* 3, 1970, pages 195–204.

"On the Packing of Ten Equal Circles In a Square." Jonathan Schaer, in *Mathematics Magazine,* 44, 1971, pages 139–140.

"Separating Points in a Rectangle." Benjamin L. Schwartz, in *Mathematics Magazine,* 46, 1973, pages 62–70.

"Configurations optimales de disques dans un polygone régulier." R. Milano. *Mémoire de Licence,* Université Libre de Bruxelles, 1987.

"A Better Packing of Ten Equal Circles In a Square." Guy Valette, in *Discrete Mathematics,* 76, 1989, pages 57–59.

"Some Progress In the Packing of Equal Circles In a Square." Michael Molland and Charles Payan, in *Discrete Mathematics,* 84, 1990, pages 303–305.

On dense packing of spheres in cubes

"On the Densest Packing of Spheres In a Cube." Jonathan Schaer, in the *Canadian Mathematical Bulletin,* 9, 1966, pages 265–270; 271–274; 275–280.

"On the Densest Packing of Equal Spheres In a Cube." Michael Goldberg, in *Mathematics Magazine,* 44, 1971, pages 199–208.

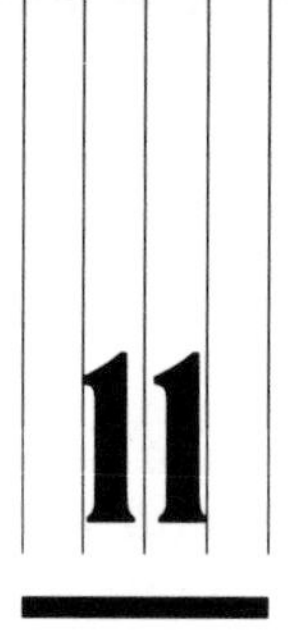

The Rotating Table and Other Problems

THE ROTATING TABLE

Imagine a square table that rotates about its center. At each corner is a deep well, and at the bottom of each well is a drinking glass that is either upright or inverted. You cannot see into the wells, but you can reach into them and feel whether a glass is turned up or down.

A move is defined as follows: Spin the table, and when it stops, put each hand into a different well. You may adjust the orientation of the glasses any way you like, that is, you may leave them as they are or turn one glass or both.

Now, spin the table again and repeat the same procedure for your second move. When the table stops spinning, there is no way to distinguish its corners, and so you have only two choices: you may reach into any diagonal pair of wells or into any adjacent pair. The object is to get all four glasses turned in the same way, either all up or all down. When this task is accomplished, a bell rings.

At the start, the glasses in the four wells are turned up or down at random. If they all happen to be turned in the same direction at this point, the bell will ring at once and the task will have been accomplished before any moves were made. Therefore it should be assumed that at the start the glasses are not all turned the same way. It is also assumed that you are not allowed to keep your hands in the wells and make experimental turnings to see if the bell rings. Furthermore, you must announce in advance the two wells to be probed at each step. You cannot probe one well, then decide which other well to probe.

Is there a procedure guaranteed to make the bell ring in a finite number of moves? Many people, after thinking briefly about this problem, conclude that there is no such procedure. It is a question of probability, they reason. With bad luck one might continue to make moves indefinitely. That is not the case, however. After no more than n correct moves one can be certain of ringing the bell. What is the minimum value of n, and what procedure is sure to make the bell ring in n or fewer moves?

Consider a table with only two corners and hence only two wells. In this case one move obviously suffices to make the bell ring. If there are three wells (at the corners of a triangular table), the following two moves suffice.

1. Reach into any pair of wells. If both glasses are turned the same way, invert both of them, and the bell will ring. If they are turned in different directions, invert the glass that is facing down. If the bell does not ring:

2. Spin the table and reach into any pair of wells. If both glasses are turned up, invert both, and the bell will ring. If they are turned in different directions, invert the glass turned down, and the bell will ring.

Although the problem can be solved in a finite number of moves when there are four wells and four glasses, it turns out that if there are five or more glasses (at the corners of tables with five or more sides), there are no procedures guaranteed to complete the task in n moves.

Turnablock

John Horton Conway's path-breaking book *On Numbers and Games* (1976) brought him a flood of correspondence suggesting new games that could be analyzed by his remarkable methods. One such suggestion, from H. W. Lenstra of Amsterdam, led Conway to develop a new family of games, one of the best of which he calls turnablock.

Turnablock is played on any n-by-n checkerboard, using n^2 counters with sides of different colors. The counters included in the board game Othello (a new name for the old British game of reversi) can be used for turnablock, or counters can be made by pairing poker chips of different colors. I shall call the two colors black and white. Here I shall describe only the simplest nontrivial version of the game, the one played on the three-by-three board.

At the start of the game the counters may be arranged in any pattern, but for the purposes of this problem assume that the game starts with the alternating-color pattern shown in Figure 76. Each player moves by turning over all the counters in any a-by-b rectangular block, where a and b are any two positive integers from 1 through 3. Thus a player's block may be a single counter, a one-by-two "domino" (oriented horizontally or vertically) or any larger configuration up to the entire three-by-three board.

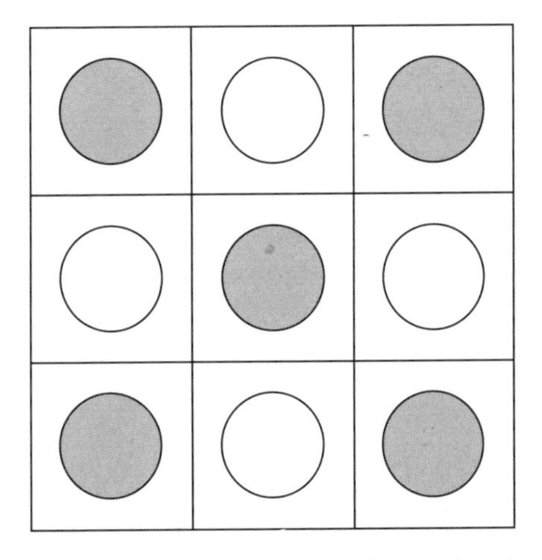

FIGURE 76 The game of turnablock

There is one essential rule in turnablock: A block may be reversed only if there is a black counter in its lower right-hand corner. It is assumed that both players are seated on the same side of the board; otherwise the player on one side may turn a block only if there is a black counter in what for him is the upper left-hand corner. The two players take turns making moves; each player must turn a block when it is his move, and the player whose move leaves all the counters with the white side up is the winner. The rule guarantees that eventually all the counters will be turned with their white side up and the game will end. With the starting pattern shown in the illustration the first player can always win if he plays correctly. What are his winning first moves, and what must his playing strategy be?

PERSISTENCES OF NUMBERS

N. J. A. Sloane of Bell Laboratories, the author of the valuable reference work *A Handbook of Integer Sequences,* introduced into number theory the concept of the "persistence" of a number. A number's persistence is the number of steps required to reduce it to a single digit by multiplying all its digits to obtain a second number, then multiplying all the digits of that number to obtain a third number, and so on until a one-digit number is obtained. For example, 77 has a persistence of four because it requires four steps to reduce it to one digit: $77 \rightarrow 49 \rightarrow 36 \rightarrow 18 \rightarrow 8$. The smallest number of persistence one is 10, the smallest of persistence two is 25, the smallest of persistence three is 39 and the smaller of persistence four is 77. What is the smallest number of persistence five?

Sloane determined by computer that no number less than 10^{50} has a persistence greater than 11. He conjectures that there is a number c such that no number has a persistence greater than c. Little is known about persistences in base notations other than 10. In base two the maximum persistence is obviously one. In base three the second term of the persistence sequence for any number is either zero or a power of 2. Sloane conjectures that in base three all powers of 2 greater than 2^{15} include a zero. Calculations show that Sloane's conjecture is true up to 2^{500}, but there is no formal proof of it. Of course, any number with zero as one of its digits is reduced to zero on the next step. Hence if the conjecture is true, it follows that the maximum persistence in base three is three, as is illustrated by the following sequence: 222,222,222, 222,222→2^{15} (which equals 1,122,221,122 in base three)→ 1,012→0. Sloane also conjectures that there is a number *c* for any base notation *b* such that no number in base *b* has a persistence greater than *c*.

Let us call Sloane's persistence a multiplicative persistence to distinguish it from additive persistence, a term introduced by Harvey J. Hindin, then a chemist at Hunter College, after he had learned of Sloane's work. The additive persistence of a number is the number of steps required to reduce it to one digit by successive additions. Recreational mathematicians and accountants know this process as "casting out nines" or obtaining a number's "digital root," procedures that are equivalent to reducing the number modulo nine. For example, 123,456,789 has an additive persistence of two: 123, 456,789→45→9.

Unlike multiplicative persistence, additive persistence is relatively trivial and almost everything about it is known. For example, in base two the smallest number of additive persistence four is 1,111,111. In base three the number is 122,222, 222. What is the smallest number of additive persistence four in base 10?

NEVERMORE

Three remarkable parodies of Edgar Allan Poe's "The Raven" appeared in issues of *Word Ways,* a fascinating journal of recreational linguistics edited and published by A. Ross Eckler (Spring Valley Road, Morristown, N.J. 07960). Each parody is based on a specific form of wordplay familiar to readers of *Word Ways.* All three poems parody the entire Poe poem, but I shall quote only the first stanzas of each.

(1)

Midnight intombed December's
naked icebound gulf.
Haggard, tired, I nodded, toiling
over my books.
Eldritch daguerreotyped dank
editions cluttered even my bed;

Exhaustion reigned.
Suddenly, now, a knocking, echoing
door I cognized:
"Eminent Boreas, open up no door!
Go, uninvited lonely frigid haunt!
Avaunt, grim guest—and roar!"

(2)

On one midnight, cold and dreary,
while I, fainting, weak and weary,
Pondered many a quaint and ancient
volume of forgotten lore,
While I studied, nearly napping,
suddenly there came a tapping,
Noise of some one gently rapping,
rapping at the chamber door.
"Oh, some visitor," I whispered,
"tapping at the chamber door,
Only one, and nothing more."

(3)

On a midnight, cool and foggy,
as I pondered, light and groggy,
Ancient books and musty ledger
not remembered any more,
As I nodded, all but napping,
there I sensed a muffled tapping,
Very much a hushful rapping,
just behind my attic door.
" 'Tis a guest, mayhap," I muttered,
"knocking at my attic door—
I can't judge it's any more."

The first parody is by Howard W. Bergerson, author of the Dover paperback *Palindromes and Anagrams*. He set himself such a difficult task that it was impossible to retain Poe's original meter and rhyme scheme. The other two parodies are by Eckler. The last was harder to write, but in both cases he was able to preserve the meter and rhyme scheme of the original. What curious linguistic structures underlie the parodies?

Rectangling the Rectangle

There is a classic dissection problem known as squaring the square: Can a square be cut into a finite number of smaller squares no two of which are the same size? The solution to this problem was discussed in an article by graph theorist William T. Tutte that is reprinted in my *2nd Scientific American Book of Mathematical Puzzles & Diversions.* At the time Tutte wrote, the best solution known ("best" meaning with a minimum number of different squares) required 24 squares. In 1978 this figure was lowered to 21 squares by A. J. W. Duijvestijn, a Dutch mathematician, as was reported in *Scientific American* (June, 1978, pages 86–88, see Figure 77). It had been known that 21 was the smallest number possible, and Duijvestijn was also able to show that his pattern for that number of squares is unique. More recently he has found two squared squares containing 22 squares and the best solution,

FIGURE 77 The lowest-order perfect square, discovered by A. J. W. Duijvestijn

also with 22 squares, for squaring the domino (a rectangle with one side twice the length of the other). See Duijvestijn's paper "A Simple Perfect Squared Square of Lowest Order," in *Journal of Combinatorial Theory,* Series B, 25, 1978, pp. 240–243.

A somewhat analogous problem is to divide a nonsquare rectangle into the minimum number of smaller rectangles in such a way that no two sides of two different rectangles have the same length. It is not hard to show that the minimum number of internal rectangles is five. Now add the condition that all sides of all six rectangles (including the outer one) are integers. Scott Kim has proved that no solution is possible in which the integers 1 through 12 are used for the twelve different edge lengths, although one can come close. Figure 78, top left, shows a solution by Kim that includes 11 twice and omits 10. If the sides of the outside rectangle are ignored, the consecutive integers from 1 through 10 will do the trick, as Ronald L. Graham has proved with the rectangle shown in Figure 78, top right.

Figure 78, bottom, is reproduced from *Mathematical Puzzles,* a challenging collection by Stephen Ainley (G. Bell and Sons, Ltd., 1977, page 59). It displays one of the two solutions in which all twelve edges are different, although not consecutive, and the area of the outside rectangle is reduced to 128. Ainley calls this a minimal area but gives no proof.

Now relax the conditions slightly so that only the ten different edges of the five internal rectangles must be distinct and integral. What solution has an outer rectangle of the smallest area? This problem was first solved by C. R. J. Singleton, who sent it to me in 1972. There are two solutions. Can you find them?

The more difficult but closely related problem of finding the smallest rectangle that can be divided into incomparable rectangles was posed by Edward M. Reingold as Problem E2422 in *The American Mathematical Monthly* (vol. 80, 1973, p. 69); a solution was published in the same journal the following year (Vol. 81, No. 6, June–July, 1974, pages 664–666). Two rectangles are called incomparable if neither one can be placed inside the other and aligned so that corresponding sides are parallel. Reingold, Andrew C. C. Yao and Bill Sands, in "Tiling with Incomparable Rectangles" (*Journal of Recreational Mathematics,* Vol. 8, 1975–76, pages 112–119), prove many theorems about this problem.

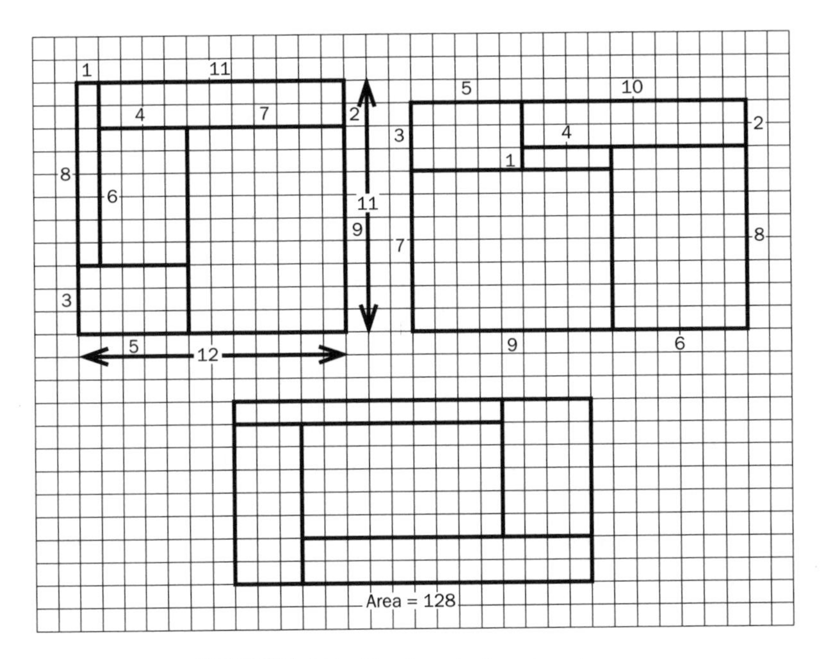

FIGURE 78 Rectangled rectangles

The minimum number of incomparable rectangles needed to tile a larger rectangle is seven. If all sides are integral, the outside rectangle with both the smallest area and the smallest perimeter is the 22-by-13 rectangle shown in Figure 79. It was found by Sands. A square can be tiled with seven incomparable rectangles having integral sides if and only if its side is 34 or larger, but eight rectangles can tile a square of side 27. This square is the smallest one known that can be tiled with incomparable rectangles, but it has not been proved to be the smallest one possible. For details consult the paper by Reingold, Yao and Sands.

In 1975 the first problem, of rectangling the rectangle, was generalized to three dimensions by Kim. There is an elegant proof that a cube cannot be cut into smaller cubes no two of which are alike. (See my *2nd Scientific American Book of Mathematical Puzzles & Diversions,* page 208.) Can a cube be "boxed" by cutting it into smaller boxes (rectangular parallelepipeds) so that no two boxes share a common edge length? The answer is yes, and Kim was able to show that the minimum number of interior boxes is 23. Later William H. Cutler

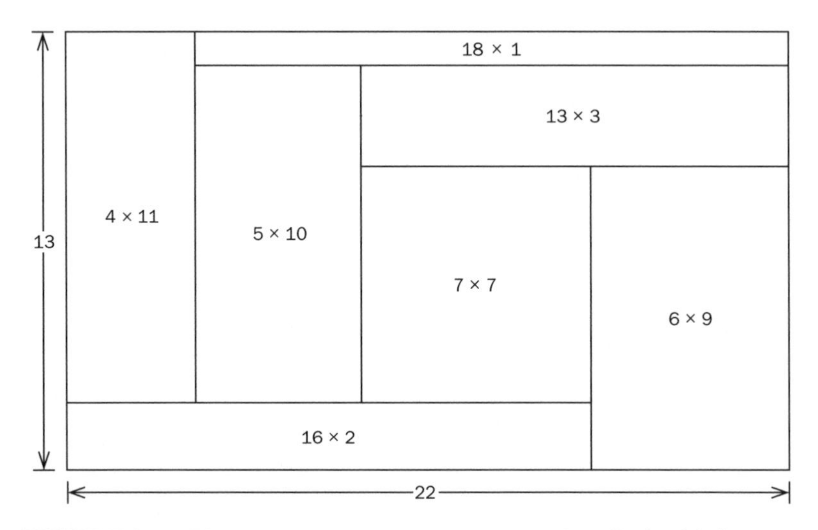

FIGURE 79 The smallest rectangle that can be tiled with incomparable rectangles

devised a second proof that 23 is minimal. Cutler found 56 essentially different ways to box the cube. If all the edges of such a cube are integral, the smallest cube that can be boxed is not known. (See "Subdividing a Box Into Completely Incongruent Boxes," by William Cutler, in *Journal of Recreational Mathematics,* 12, 1979–1980, pp. 104–111.)

Another unsolved problem is to determine the noncubical box of smallest volume that can be sliced into 23 (or possibly more) boxes with no edge in common and all edges integral. Cutler found a box that is 147 by 157 by 175 and splits into the following 23 boxes:

13 by 112 by 141	27 by 36 by 48
18 by 72 by 82	34 by 110 by 135
23 by 41 by 73	57 by 87 by 97
31 by 69 by 78	16 by 74 by 140
38 by 42 by 90	21 by 52 by 65
14 by 70 by 75	28 by 55 by 123
19 by 53 by 86	35 by 62 by 127
26 by 49 by 56	17 by 24 by 67
33 by 46 by 60	22 by 107 by 131
45 by 68 by 85	30 by 54 by 134
15 by 44 by 50	37 by 83 by 121
20 by 40 by 92	

It is not easy to fit these boxes together to make the large box. As far as I know no one has worked on a three-dimensional version of Reingold's incomparable rectangles.

Three Geometric Puzzles

1. The puzzle shown in Figure 80 is reproduced from the September–October, 1978 issue of the magazine *Games.* The task is to trace in the larger figure a shape geometrically similar to the smaller one shown below it.

2. The puzzle shown in Figure 81 is from a special issue of the French magazine *Science et Vie* (September, 1978) that was devoted entirely to recreational mathematics. In each row the third pattern is obtained from the first two by applying a rule. What is the rule, and what pattern goes in the blank space in the third row?

3. The trapezoid shown in Figure 82 is called a triamond, or an order-three polyiamond, because it can be formed by joining three equilateral triangles. In a show last fall at the 55 Mercer Gallery in New York, Denis McCarthy exhibited a striking tessellation made up of 174 of these shapes. They were cut from corrugated cardboard, so that under slanting light their ribbed surfaces would create patterns of light and dark triamonds that would vary with the position of the viewer [see Figure 83].

There is an old puzzle that asks how to cut the triamond into four congruent parts. Figure 82 gives the traditional solution. Richard Brady, a mathematics teacher in Washington, D.C., tells me that when Andrew Miller, one of his pupils, encountered the problem in Harold R. Jacobs' *Geometry* (W. H. Freeman and Company, 1974, page 188), he found a different solution. In Miller's new solution all four regions do not have the same shape as the larger figure, but they are identical if one or more may be turned over. What is the new solution?

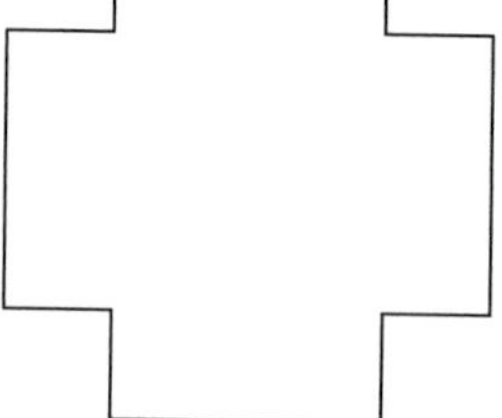

FIGURE 80 Trace the shape of the smaller figure in the larger one

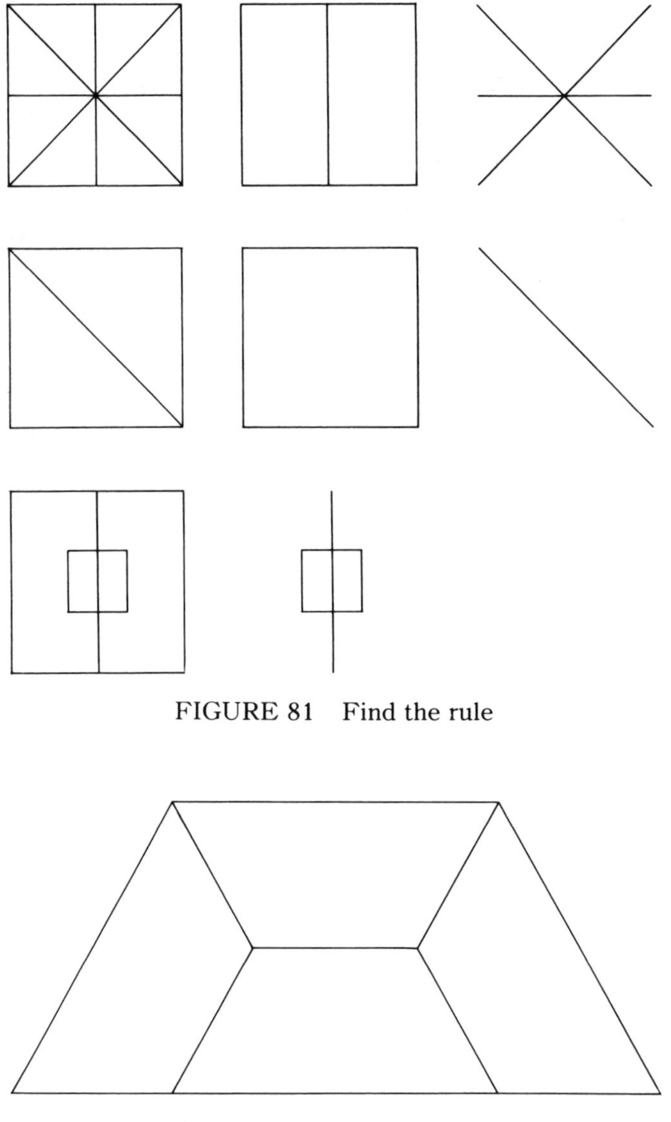

FIGURE 81 Find the rule

FIGURE 82 Do it another way

FIGURE 83 Detail of Denis McCarthy's tessellation of 174 triamonds

ADDENDUM

I had confined rectangling to non-squares. If the outer "rectangle" is allowed to be square, it can be cut into five rectangles for which the ten different sides have the lengths 1 through 10, and the outside square has a side of 11. This pretty pattern, discovered by M. den Hertog, of Belgium, is shown in Figure 84.

I said I knew of no work on incomparable tiling of cuboids (rectangular parallelepipeds) by smaller cuboids—that is, a tiling such that given any pair of the boxes, neither will fit inside the other. Just such work was discussed by Richard Guy in his note on "How Few Incomparable Cuboids Will Tile a Cube?" (*American Mathematical Monthly,* 91, 1984, pp. 624–629.) The smallest number of incomparable cuboids that will tile a cube is known to be at least seven, and the smallest number that will tile a noncubical cuboid is six.

In a later note on "Unsolved Problems," (*Ibid,* 92, 1985, pp. 725–732) Guy reports on several remarkable discoveries

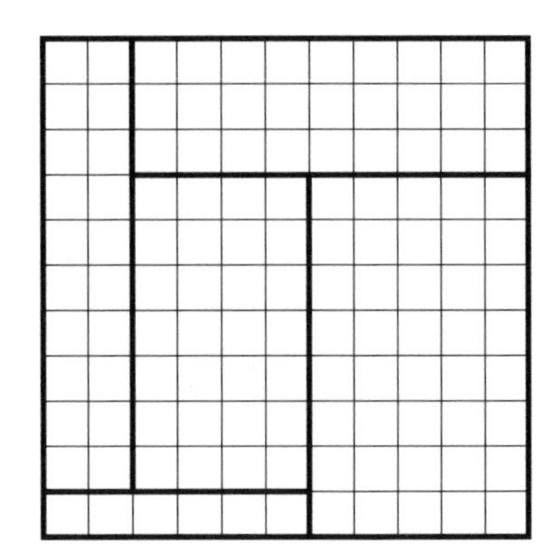

FIGURE 84 A rectangled square of sides 1 through 11, and area of 121

by Charles H. Jepsen, of Grinell College. Jepsen covers them at greater length in his paper "Tiling with Incomparable Cuboids," in *Journal of Recreational Mathematics,* 59, 1986, pp. 283–292. He proves that no incomparable tiling of a cuboid can have fewer than six pieces, and gives an example of such a tiling of a $3 \times 5 \times 9$ cuboid, along with a proof that its volume of 135 is minimal. He also proves that no tiling is possible of a $3 \times 3 \times C$ cuboid.

Jepsen also gives an incomparable tiling of a $10 \times 10 \times 10$ cube with seven pieces, and conjectures that its 10^3 volume is minimal. The paper closes with four unanswered questions.

ANSWERS

THE ROTATING TABLE

The following five moves will suffice to get all four glasses in the wells of the rotating table turned either all up or all down.

1. Reach into any diagonal pair of wells. If the glasses are not both turned up, adjust them so that they are. If the bell does not ring:

2. Spin the table and reach into any adjacent pair of wells. If both glasses are turned up, leave them that way; otherwise invert the glass that is turned down. If the bell fails to ring, you know that now three glasses are turned up and one is turned down.

3. Spin the table and reach into any diagonal pair of wells. If one of the glasses is turned down, invert it and the bell will ring. If both are turned up, invert one so that the glasses are arranged in the following pattern:

Up Down
Up Down

4. Spin the table and reach into any adjacent pair of wells. Invert both glasses. If they were both turned in the same direction, the bell will ring; otherwise the glasses are now arranged in the following pattern:

Up Down
Down Up

5. Spin the table, reach into any diagonal pair of wells and invert both glasses. The bell will ring.

Ronald L. Graham of Bell Laboratories and Persi Diaconis of Harvard University have together considered two ways of generalizing the problem. One is to assume that the player has more than two hands. If the number of glasses n is greater than 4, the problem is solvable with $n-2$ hands if and only if n is not a prime number. Therefore for five glasses (5 is a prime) the problem is not solvable with $5-2$, or 3, hands. It may be that if n is a composite number, then for some values of n the problem can be solved with fewer than $n-2$ hands. The problem can also be generalized by replacing glasses with objects that have more than two positions.

Ted Lewis and Stephen Willard, writing on "The Rotating Table" in *Mathematics Magazine* (53, 1980, pp. 174–175) were the first to publish a solution for the general case involving a rotating polygonal table with n wells, and a player with k hands whom they called a "bell ringing octopus." They showed that the player can always force the bell to ring in a finite number of steps if and only if k is equal or greater than $(1-1/p)n$, where p is the largest prime factor of n.

This result was also reached independently by a number of readers. Unless I have misplaced some letters they were

Lyle Ramshaw, Richard Litherland, Eugene Gover and Nishan Krikorian, Richard Ahrens and John Mason, and Sven Eldring. The most detailed proof, along with related results and conjectures, can be found in "Probing the Rotating Table," by Ramshaw and William T. Laaser, in *The Mathematical Gardner,* edited by David A. Klarner (Wadsworth, 1981, pp. 288–307).

Two amusing robot variations were proposed by Albert G. Stanger as Problem 1598, "Variations on the Rotating Table Problem," in the *Journal of Recreational Mathematics* (19, 1987, pp. 307–308).

1. Instead of reaching into the wells, indicate to a robot the two you want probed. The robot does the reaching, then says either "Same" (if both glasses have the same orientation), or "Different." You then command the robot either to flip no glass, flip both, or flip only one which he selects at random.

2. This is the same as the previous variant except that the robot may tell a lie at any time.

In his solution (*Ibid,* 20, 1988, pp. 312–314), Stanger showed that in both variants the bell can be made to ring in a finite number of steps. He gives a five-step solution for the first variation, and a seven-step solution for the second. In the second strategy, it makes no difference if the robot always tells the truth, always lies, or mixes truths and lies, because the strategy pays no attention to what the robot says! This surprising strategy, in which you pay no attention to anything, provides a seven-step solution to the original problem. It had been sent to me earlier in a March 1979 letter from Miner S. Keeler, president of the Keeler Brass Company in Grand Rapids, Michigan, who expressed surprise that I had not mentioned it when I answered the original problem. Here is how it works:

1. Invert any diagonal pair.
2. Invert any adjacent pair.
3. Invert any diagonal pair.
4. Invert any single glass.
5. Invert any diagonal pair.
6. Invert any adjacent pair.
7. Invert any diagonal pair.

If you don't believe these steps are certain to ring the bell after at most seven steps, try it out by placing four playing cards on the table in any pattern of up and down. Regardless of the initial pattern, you'll find that the strategy will produce either four up cards or four down cards in at most seven moves.

In a note added to Stanger's answer to his problem, Douglas J. Lanska generalized the robot variation to p wells and q hands. He pointed out that if the robot always lies or always tells the truth it is not necessary to know which is the case. Simply follow the n-step strategy on the assumption that the robot is a truthteller. If the bell fails to ring after n steps, you know the robot is a liar. Then repeat the same n steps, but now reverse the robot's answers. The bell will ring in another n steps, or $2n$ steps in all.

Gover and Krikorian, in a later letter, reported some results on rotating cubes and other regular polyhedra.

I first heard of the original problem from Robert Tappay, a Toronto friend, who was told of it by a mathematician at the University of Quebec. He in turn had found it in a test given to Russian mathematics students. Someone in Poland, he told Tappay, had written his doctor's thesis on how people go about trying to solve the problem.

TURNABLOCK

The first player's strategy for winning John Horton Conway's order-three game of turnablock requires numbering the nine cells as is shown at the top of Figure 85. The player must only make moves that leave black counters on cells whose numbers have a "nim sum" of zero. The nim sum of a collection of numbers is traditionally obtained by writing the numbers in binary form and adding without carrying. If all the digits in the total are zero, then the nim sum is zero. Conway suggests a simpler procedure for finding a nim sum: express each of the numbers to be added as a sum of distinct powers of 2, and then cancel pairs of like powers. If no powers remain, then the nim sum is zero; otherwise the nim sum is simply the sum of the remaining powers. For example, consider $1+5+12$. Writing each number in this expression as the sum of distinct powers of 2 gives $1+4+1+8+4$. The pairs of 1's and 4's cancel, leaving a nim sum of 8. Similarly, consider

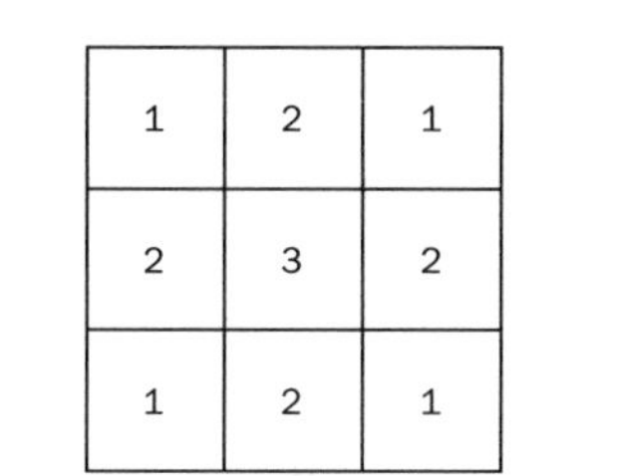

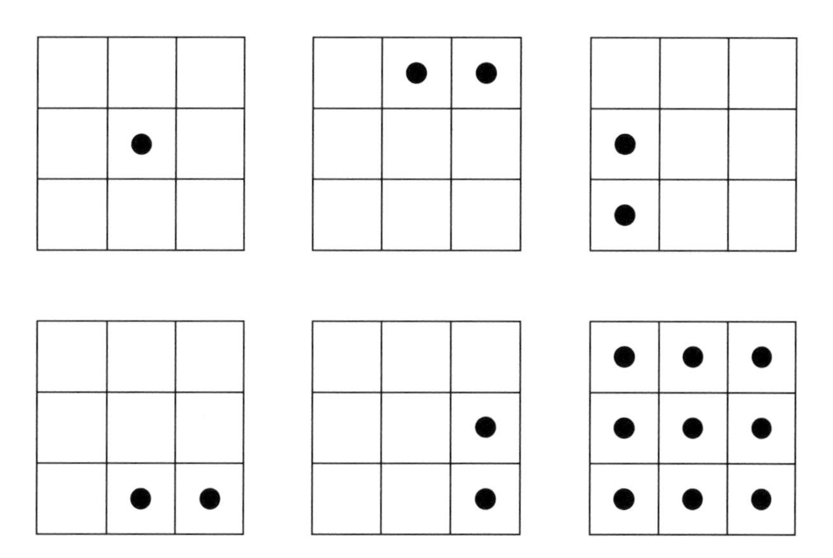

FIGURE 85 Winning strategies for three-by-three turnablock

$1+2+5+6$. This expression becomes $1+2+4+1+4+2$. All the powers cancel, and the nim sum is zero.

The dots in Figure 85 indicate the blocks that can be turned over for a winning first move. In other words, after each of these moves black counters are left on cells whose numbers have a nim sum of zero. Any second move by the other player will necessarily leave black counters on cells with a nonzero nim sum. The first player will always be able to respond with a zero nim-sum move, and by continuing in this way he is sure to win.

Figure 86, top, supplied by Conway, gives the cell numbering for all rectangular turnablock boards with sides from one through eight. The numbering of an *a*-by-*b* board smaller

1	2	1	4	1	2	1	8
2	3	2	8	2	3	2	12
1	2	1	4	1	2	1	8
4	8	4	6	4	8	4	11
1	2	1	4	1	2	1	8
2	3	2	8	2	3	2	12
1	2	1	4	1	2	1	8
8	12	8	11	8	12	8	13

Persistence	Number
1	10
2	25
3	39
4	77
5	679
6	6,788
7	68,889
8	2,677,889
9	26,888,999
10	3,778,888,999
11	277,777,788,888,899

FIGURE 86 Strategy chart for turnablock (left). Smallest numbers with persistence 11 or less (right).

than the order-eight board is given by the a-by-b rectangle in the upper left-hand corner of the matrix. The numbers in the cells are obtained by a process called nim multiplication, which you will find explained in Conway's *On Numbers and Games* (Academic Press, 1976, page 52).

PERSISTENCIES OF NUMBERS

The smallest number in base 10 of multiplicative persistence 5 is 679. The chart in Figure 86, bottom, gives the smallest numbers with multiplicative persistence of 11 or less. It is taken from N. J. A. Sloane's paper "The Persistence of a Number," in *Journal of Recreational Mathematics* (6, 1973, pp. 97–98).

The smallest number in base 10 of additive persistence 4 is 19,999,999,999,999,999,999,999. The sum of the digits in this number is 199, the smallest number of additive persistence 3. The sum of the digits in 199 is 19, the smallest number of additive persistence 2. The sum of the digits in 19 is 10, the smallest number of additive persistence 1. More generally, the second step in the sequence for the smallest number of additive persistence k gives the smallest number of additive persistence $k-1$. All such numbers start with 1 and are followed by 9's. Therefore the smallest number of additive persistence 5 is the number consisting of 1 followed by

2,222,222,222,222,222,222,222 9's. For further discussion of the problem see Harvey J. Hinden's "The Additive Persistence of a Number" (*Journal of Recreational Mathematics,* 7, 1974, pp. 134–135).

NEVERMORE

The first parody of Edgar Allan Poe's "The Raven" is by Howard W. Bergerson, and is called an automynorcagram. The first letters of each word in the parody, taken in the order they appear, spell out the first verse of the parody. For the complete parody see *Word Ways* (8, 1975, pp. 219–222).

The second parody of Poe's poem is by A. Ross Eckler, editor and publisher of *Word Ways*. It is homoliteral, that is, each consecutive pair of words have at least one letter in common. The full parody appears in *Word Ways* (9, 1976, pp. 96–98).

The third parody, also by Eckler, is found in *Word Ways* (9, 1976, pp. 231–233; November, 1976). It is heteroliteral, that is, each pair of consecutive words has no letter in common.

A well known American poet who asked me not to mention his name took issue with my statement that it is not possible to write a parody of "The Raven" in which the initial letters of each word spells out the original stanzas, and at the same time retain Poe's rhyme and meter scheme. Here is how he managed it:

One November evening, nodding over volumes ever-plodding,
My bedeviled eyes regretting every volume, evermore
Nincompoopishly I noodled, garnered nonsense-rhymes, or doodled,
Doing interlineations nine-times gerrymandered o'er:
Vain, elusive ruminations, variations on "Lenore,"
Uttering mutterings encore.

Shh! Egad! Verandah-tapping! Ectoplasmical rap-rapping
Preternaturally lapping one's dilapidated door!
Inchwise noise gone madly yowling by each dulled ear, vampire-howling
Implings licking every doorway, every yielding entrydoor,
Shades ranged everywhere, grim rappers ever tougher to ignore,
Noisily gnashing extempore.

"Visitors!" exclaimed, rewoken, Yours Voluminously. "Oaken
Laminationwork unbroken, metalclad exterior,
Even vault-thick engine-metal reinforcement may outfettle

Regulation Elementals, not Immeasurables nor
Cosmic Omnicompetences. My poor ostel's open. Pour
In, sweet heebiejeebs, like yore."

RECTANGLING THE RECTANGLE

I originally said that Singleton's problem had only one solution, and the one I gave was incorrectly drawn. In 1979 I received a letter from M. den Hertog, containing much interesting material on rectangling theory. He showed that for every rectangled rectangle one could obtain another pattern by using the same side lengths. The two solutions shown in Figure 87 appeared in the *Journal of Recreational Mathematics* (12, 1979–80, pp. 147–148) as a solution by Jean Meeus to Problem 731.

THREE GEOMETRIC PUZZLES

1. The solution is shown in Figure 88. The puzzle is adapted from an optical illusion in Charles H. Paraquin's *Eye Teasers; Optical Illusion Puzzles* (Sterling Publishing Co., Inc., 1978).

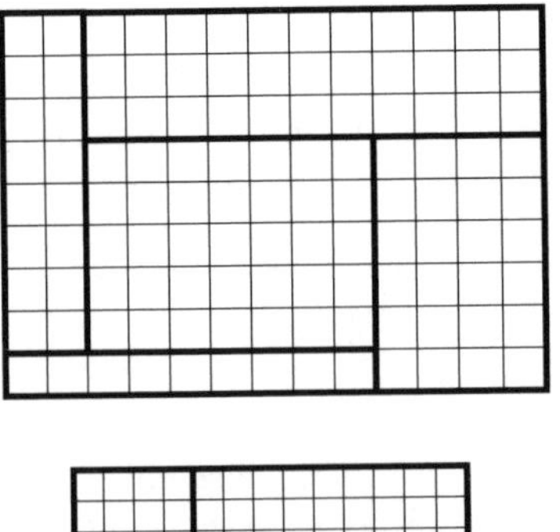

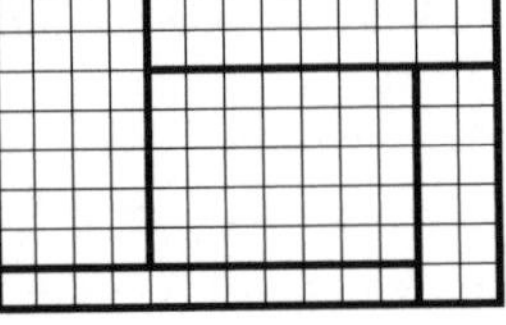

FIGURE 87 The only two solutions to Singleton's puzzle

FIGURE 88 Solution to a problem

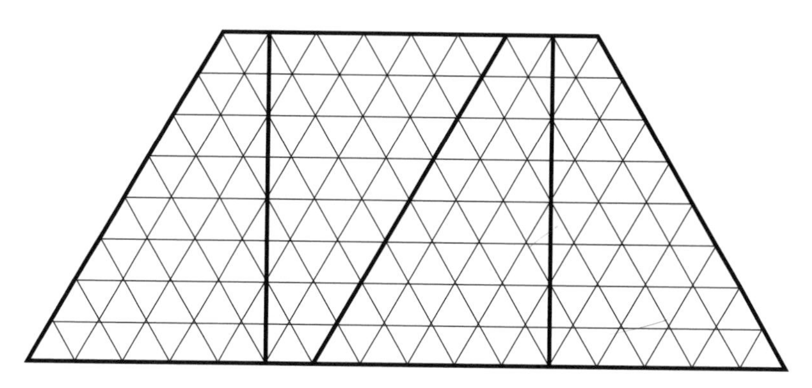

FIGURE 89 Triamond divided into four identical regions

2. The rule for obtaining the third pattern in each row is to superpose the first two patterns and eliminate any lines they have in common. Hence the pattern to be placed at the end of the third row is simply a square.

3. Figure 89 displays the only other way the triamond can be cut into four identical regions.

12

Does Time Ever Stop? Can the Past Be Altered?

"It is impossible to meditate on time and the mystery of the creative passage of nature without an overwhelming emotion at the limitations of human intelligence."

—ALFRED NORTH WHITEHEAD,
The Concept of Nature

There has been a great deal of interest among physicists of late in whether or not there are events on the elementary-particle level that cannot be time-reversed, that is, events for which imagining a reversal in the direction of motion of all the particles involved is imagining an event that cannot happen in nature. Richard Feynman has suggested an approach to quantum mechanics in which antiparticles are viewed as particles momentarily traveling backward in time. Cosmologists have speculated about two universes for which all the events in one are reversed relative to the direction of time in the other: in each universe intelligent organisms would live normally from past to future, but if the organisms in one universe could in some way observe events in the other (which many physicists consider an impossibility), they would find those events going in the opposite direction. It has even been conjectured that if our universe stops expanding and starts to contract, there will be a time reversal, but it is far from clear what that would mean. Most of the speculations of this kind are quite recent, and interested readers will find many of them examined in four chapters of my *New Ambidextrous Universe.*

In this chapter I shall consider two bizarre questions about time that are not discussed in the book. Indeed, these questions are of so little concern to scientists that only philosophers and writers of fantasy and science fiction have had much to say about them: Is it meaningful to speak of time stopping? Is it meaningful to speak of altering the past?

Neither question should be confused with the familiar subject of time's relativity. Newton believed the universe was pervaded by a single absolute time that could be symbolized by an imaginary clock off somewhere in space (perhaps outside the cosmos). By means of this clock the rates of all the events in the universe could be measured. The notion works well within a single inertial frame of reference such as the surface of the earth, but it does not work for inertial systems moving in relation to each other at high speeds. According to the theory of relativity, if a spaceship were to travel from our solar system to another solar system with a velocity close to that of light, events would proceed much slower on the spaceship than they would on the earth. In a sense, then, such a spaceship is traveling through time into the future. Passengers on the spaceship might experience a round-trip voyage as taking only a few years, but they would return to find that centuries of earth-years had elapsed.

The notion that different parts of the universe can change at different rates of time is much older than the theory of relativity. In the Scholastic theology of the Middle Ages angels were considered to be nonmaterial intelligences living by a time different from that of earthly creatures; God himself was thought to be entirely outside of time. In the first act of Lord Byron's play, *Cain, A Mystery,* the fallen angel Lucifer says:

With us acts are exempt from time.
and we
Can crowd eternity into an hour
Or stretch an hour into eternity
We breathe not by a mortal
measurement—
But that's a mystery.

In the 20th century hundreds of science-fiction stories have played with the relativity of time in different inertial systems, but the view that time can speed up or slow down in different parts of our universe is central to many older tales. A popular medieval legend tells of a monk who is entranced for a minute or two by the song of a magical bird. When the bird stops singing, the monk discovers that several hundred years have passed. In a Moslem legend Mohammed is carried by a mare into the seventh heaven. After a long visit the prophet returns to the earth just in time to catch a jar of water the horse had kicked over before starting its ascent.

Washington Irving's "Rip Van Winkle" is this country's best-known story about someone who sleeps for what seems to him to be a normal time while two decades of earth-years rush by. King Arthur's daughter Gyneth slept for 500 years under a spell cast by Merlin. Every culture has similar sleeper legends. H. G. Wells used the device in *When the Sleeper Wakes,* and it is a common practice in science fiction to put astronauts into a cryogenic sleep so they can survive interstellar voyages that are longer than their normal life span. In Wells's short story "The New Accelerator" a scientist discovers a way to speed up a person's biological time so that the world seems to come almost to a halt. This device too is frequently encountered in later science fiction.

The issue under consideration here, however, is not how time can vary but whether time can be said to stop entirely. It is clearly meaningful to speak of all motion ceasing in one part of the universe, whether or not such a part exists. In the theory of relativity the speed of light is an unattainable limit for any object with mass. If a spaceship could attain the speed of light (which the theory of relativity rules out because the mass of the ship would increase to infinity), then time on the spaceship would stop in the sense that all change on it would cease. In earth time it might take 100 years for the spaceship to reach a destination, but to astronauts on the spaceship the destination would be reached instantaneously. One can also imagine a piece of matter or even a human being reduced to such a low temperature (by some as yet unknown means) that even all subatomic motions would be halted. For that piece of matter, then, one could say that time had stopped. Actually it is hard to understand why the piece of matter would not vanish.

The idea of time stopping creates no problems for writers of fantasy, who are not constrained by the real world. For example, in L. Frank Baum's "The Capture of Father Time," one of the stories in his *American Fairy Tales* (now back in print in a Dover facsimile edition), a small boy lassoes Time, and for a while everything except the movements of the boy and Father Time stops completely. In Chapter 22 of James Branch Cabell's *Jurgen: A Comedy of Justice,* outside time sleeps while Jurgen enjoys a pleasurable stay in Cocaigne with Queen Anaïtis. Later in the novel Jurgen stares into the eyes of the God of his grandmother and is absolutely motionless for 37 days. In Jorge Luis Borges' story "The Secret Miracle" a writer is executed by a firing squad. Between the command to fire and the writer's death God stops all time outside the writer's brain, giving him a year to complete his masterpiece.

Many similar examples from legend and literature show that the notion of time stopping in some part of the universe is not logically inconsistent. But what about the idea of time stopping throughout the universe? Does the notion that everything stops moving for a while and then starts again have any meaning?

If it is assumed that there is an outside observer—perhaps a god—watching the universe from a region of hypertime, then of course the notion of time stopping does have meaning, just as imagining a god in hyperspace gives meaning to the notion of everything in the universe turning upside down. The history of our universe may be like a three-dimensional motion picture a god is enjoying. When the god turns off the projector to do something else, a few millenniums may go by before he comes back and turns it on again. (After all, what are a few millenniums to a god?) For all we can know a billion centuries of hypertime may have elapsed between my typing the first and the second word of this sentence.

Suppose, however, all outside observers are ruled out and "universe" is taken to mean "everything there is." Is there still a way to give a meaning to the idea of all change stopping for a while? Although most philosophers and scientists would say there is not, a few have argued for the other side. For example, in "Time without Change" Sydney S. Shoemaker, now a philosopher at Cornell University, makes an unusual argument in support of the possibility of change stopping.

Shoemaker is concerned not with the real world but with possible worlds designed to prove that the notion of time stopping everywhere can be given a reasonable meaning. He proposes several worlds of this kind, all of them based on the same idea. I shall describe only one such world here, in a slightly dramatized form.

Imagine a universe divided into regions *A, B* and *C*. In normal times inhabitants of each region can observe the inhabitants of the other two and communicate with them. Every now and then, however, a mysterious purple glow permeates one of the regions. The glow always lasts for a week and is invariably followed by a year in which all change in the region ceases. In other words, for one year absolutely nothing happens there. Shoemaker calls the phenomenon a local freeze. Since no events take place, light cannot leave the region, and so the region seems to vanish for a year. When it returns to view, its inhabitants are unaware of any passage of time, but they learn from their neighbors that a year, as measured by clocks in the other two regions, has elapsed. To the inhabitants of the region that experienced the local freeze it seems that instantaneous changes have taken place in the other two regions. As Shoemaker puts it: "People and objects will appear to have moved in a discontinuous manner or to have vanished into thin air or to have materialized out of thin air; saplings will appear to have grown instantaneously into mature trees, and so on."

In the history of each of the three regions local freezes, invariably preceded by a week of purple light, have happened thousands of times. Now suppose that suddenly, for the first time in history, purple light appears simultaneously in regions *A, B* and *C* and lasts for a week. Would it not be reasonable, Shoemaker asks, for scientists in the three regions to conclude that change had ceased for a year throughout the entire universe even though no minds were aware of it?

Shoemaker considers several objections to his thesis and counters all of them ingeniously. Interested readers can consult his paper and then read a technical analysis of it in the fifth chapter of G. Schlesinger's *Confirmation and Confirmability*. Schlesinger agrees with Shoemaker that an empirical, logically consistent meaning can be found for the sentence "A period of time *t* has passed during which absolutely nothing happened." Note that similar arguments about possible worlds can provide meanings for such notions as everything in a universe turning upside down, mirror-reversing, doubling in size and so on.

The question of whether the past can be changed is even stranger than that of whether time can stop. Writers have often speculated about what might have happened if the past had taken a different turn. J. B. Priestley's play *Dangerous Corner* dealt with this question, and there have been innumerable "what if" stories in both science fiction and other kinds of literature. In all time-travel stories where someone enters the past, the past is necessarily altered. The only way the logical contradictions created by such a premise can be resolved is by positing a universe that splits into separate branches the instant the past is entered. In other words, while time in the old branch "gurgles on" (a phrase from Emily Dickinson) time in the new branch gurgles on in a different way toward a different future. When I speak of altering the past, however, I mean altering it throughout a single universe with no forking time paths. (Pseudoalterations of the past, such as the rewriting of history satirized by George Orwell in *1984,* obviously do not qualify.) Given this context, can an event, once it has happened, ever be made not to have happened?

The question is older than Aristotle, who in his *Ethics* (Book 6) writes: "It is to be noted that nothing that is past is an object of choice, for example, no one chooses to have sacked Troy; for no one *deliberates* about the past, but about what is future and capable of being otherwise, while what is past is not capable of not having taken place; hence Agathon is right in saying: 'For this alone is lacking even to God, to make undone things that have once been done.' "

Thomas Aquinas believed God to be outside of time and thus capable of seeing all his creation's past and future in one blinding instant. (Even though human beings have genuine power of choice, God knows how each one will choose; it is in this way that Aquinas sought to harmonize predestination and free will.) For Aquinas it was not possible for God to do absolutely impossible things, namely those that involve logical contradiction. For example, God could not make a creature that was both a human being and a horse (that is, a complete human being and a complete horse, rather than a mythical combination of parts such as a centaur), because that would involve the contradiction of assuming a creature to be simultaneously rational and nonrational.

Similarly, God cannot alter the past. That would be the same as asserting that the sack of Troy both took place and did not take place. Aquinas agreed with Aristotle that the past must forever be what it was, and it was this view that became the official position of medieval Scholasticism. It is not so much that God's omnipotence is limited by the law of contradiction but rather that the law is part of God's nature. "It is best to say," Aquinas wrote, "that what involves contradiction cannot be done rather than that God cannot do it." Modern philosophers would say it this way. God can't make a four-sided triangle, not because he can't make objects with four sides but because a triangle is *defined* as a three-sided polygon. The phrase "four-sided triangle" is therefore a nonsense phrase, one without meaning.

Edwyn Bevan, in a discussion of time in his book *Symbolism and Belief,* finds it odd that Aquinas would deny God the ability to alter the past and at the same time allow God to alter the future. In the 10th question of *Summa Theologica* (Ia. 10, article 5.3), Aquinas wrote: "God can cause an angel not to exist in the future, even if he cannot cause it not to exist while it exists, or not to have existed when it already has." For Aquinas to have suggested that for God the past is unalterable and the future is not unalterable, Bevan reasons, is surely to place God in some kind of time, thus contradicting the assertion that God is outside of time.

I know of no scientist or secular philosopher who has seriously believed the past could be altered, but a small minority of theologians have maintained that it could be. The greatest of them was Peter Damian, the zealous Italian reformer of the Roman Catholic church in the 11th century. In *On Divine Omnipotence,* his most controversial treatise, Damian argued that God is in no way bound by the law of contradiction, that his omnipotence gives him the power to do all contradictory things including changing the past. Although Damian, who started out as a hermit monk, argued his extreme views skillfully, he regarded all reasoning as superfluous, useful only for supporting revealed theology. It appears that he, like Lewis Carroll's White Queen, would have defended everyone's right to believe six impossible things before breakfast. (Damian was also a great promoter of self-flagellation as a form of penance, a practice that became such a fad during his lifetime that some monks flogged themselves to death.)

One of my favorite Lord Dunsany stories is the best example I know of from the literature of fantasy that illustrates Damian's belief in the possibility of altering the past. It is titled "The King That Was Not," and you will find it in Dunsany's early book of wonder tales *Time and the Gods*. It begins as follows: "The land of Runazar hath no King nor ever had one; and this is the law of the land of Runazar that, seeing that it hath never had a King, it shall not have one for ever. Therefore in Runazar the priests hold sway, who tell the people that never in Runazar hath there been a King."

The start of the second paragraph is surprising: "Althazar, King of Runazar. . . ." The story goes on to recount how Althazar ordered his sculptors to carve marble statues of the gods. His command was obeyed, but when the great statues were undraped, their faces were very much like the face of the king. Althazar was pleased and rewarded his sculptors handsomely with gold, but up in Pegāna (Dunsany's Mount Olympus) the gods were outraged. One of them, Mung, leaned forward to make his sign against Althazar, but the other gods stopped him: "Slay him not, for it is not enough that Althazar shall die, who hath made the faces of the gods to be like the faces of men, but he must not even have ever been."

"Spake we of Althazar, a King?"
asked one of the gods.
"Nay, we spake not."
"Dreamed we of one Althazar?"
"Nay, we dreamed not."

Below Pegāna, in the royal palace, Althazar suddenly passed out of the memory of the gods and so "became no longer a thing that was or had ever been." When the priests and the people entered the throne room, they found only a robe and a crown. "The gods have cast away the fragment of a garment," said the priests, "and lo! from the fingers of the gods hath slipped one little ring."

ADDENDUM

When I wrote about time stopping I was using a colloquial expression to mean that change ceases. Because there are no moving "clocks" of any sort for measuring time, one can say in a loose sense that time stops. Of course time does not move or stop any more than length can extend or not extend. It is the universe that moves. You can refute the notion that time "flows" like a river simply by asking: "At what rate does it flow?" Shoemaker wanted to show in his paper not that time stops, then starts again, but that all change can stop and some sort of transcendental hypertime still persists. Change requires time, but perhaps, Shoemaker argued, time does not require change in our universe.

Harold A. Segal, in a letter in the *New York Times* (January 11, 1987) quoted a marvelous passage from Shakespeare's *As You Like It* (Act III, Scene 2) in which Rosalind explains how time can amble, trot, gallop, or stand still for different persons in different circumstances. It trots for the "young maid" between her engagement and marriage. It ambles for a priest who knows no Latin because he is free from the burden of "wasteful learning." It ambles for the rich man in good health who "lives merrily because he feels no pain." It gallops for the thief who awaits his hanging. For whom does it stand still? "With lawyers in the vacation; for they sleep between term and term, and then they perceive not how Time moves."

Isaac Asimov, in an editorial in *Asimov's Science-Fiction Magazine* (June, 1986) explained why it would not be possible for a person to walk about and observe a world in which all change had stopped. To move, she would have to push aside molecules, and this would inject time into the outside world. She would be as frozen as the universe, even though dancing atoms in her brain might continue to let her think. Asimov could have added that she would not even be able to *see* the world because sight depends on photons speeding from the world into one's eyes.

Two readers, Edward Adams and Henry Lambert, independently wrote to say that the god Koschei, in *Jurgen,* could alter the past. At the end of the novel he eliminates all of Jurgen's adventures as never having happened. However, Jurgen recalls that Horvendile (the name Cabell often used for himself) once told him that he (Horvendile) and Koschei were one and the same!

Edward Fredkin is a computer scientist who likes to think of the universe as a vast cellular automaton run by an inconceivably complex algorithm that tells the universe how to jump constantly from one state to the next. Whoever or whatever is running the program could, of course, shut it down at any time, then later start it running again. We who are part of the program would have no awareness of such gaps in time.

On the unalterability of the past, readers reminded me of the stanza in Omar's *Rubaiyat* about the moving finger that having writ moves on, and all our piety and wit cannot call it back to cancel half a line. Or as Ogden Nash once put it:

One thing about the past.
It's likely to last.

I touched only briefly on the many science-fiction stories and novels that deal with time slowing down or halting. For references on some of the major tales see the section "When time stands still" on page 153 of *The Visual Encyclopedia of Science Fiction*. The most startling possibility, now seriously advanced by some physicists, is that the universe comes to a complete stop billions of times every microsecond, then starts up again. Like a cellular automaton it jumps from state to state. Between the jumps, nothing changes. The universe simply does not exist. Time is quantized. An electron doesn't move smoothly from here to there. It moves in tiny jumps, occupying no space in between.

The fundamental unit of quantized time has been called the "chronon." Between chronons one can imagine one or more parallel universes operating within our space, but totally unknown to us. Think of a film with two unrelated motion pictures running on alternate frames. Between the frames of our universe, who knows what other exotic worlds are unrolling in the intervals between our chronons? Both motion pictures and cellular automata are deterministic, but in this vision of parallel universes running in the same space, there is no need to assume determinism. Chance and free will could still play creative roles in making the future of each universe unpredictable in principle.

BIBLIOGRAPHY

"Seven Sleepers" and "Sleepers". E. Cobham Brewer, in *The Reader's Handbook,* revised edition. Lippincott, 1898.

"Time." C. D. Broad, in *Encyclopedia of Religion and Ethics,* James Hastings (ed.). Scribner's, 1922.

Travelers in Time: Strange Tales of Man's Journeying into the Past and the Future. Philip Van Doren Stern (ed.). Doubleday, 1947.

Symbolism and Belief. Edwyn Bevan. Kennikar Press, 1968.

"Time Without Change." Sidney S. Shoemaker, in *The Journal of Philosophy,* 66, 1969, pages 363–381.

Confirmation and Confirmability. G. Schlesinger. Oxford University Press, 1974.

"Time and Nth Dimensions." Kenneth Bulmer, in *The Visual Encyclopedia of Science Fiction.* Harmony Books, 1977, pages 145–154.

"Time Travel." Anonymous, in *The Science Fiction Encyclopedia.* Peter Nicholls (ed.). Doubleday, 1979.

"Time Travel and Other Universes." Anonymous, in *The Science in Science Fiction,* Chapter 5. Peter Nicholls (ed.) Knopf, 1983.

"Time Travel." Martin Gardner, in *Time Travel and Other Mathematical Bewilderments.* W. H. Freeman, 1988.

The New Ambidextrous Universe. Martin Gardner. W. H. Freeman, 1990.

13

Generalized Ticktacktoe

The world's simplest, oldest and most popular pencil-and-paper game is still ticktacktoe, and combinatorial mathematicians, often with the aid of computers, continue to explore unusual variations and generalizations of it. In one variant that goes back to ancient times the two players are each given three counters, and they take turns first placing them on the three-by-three board and then moving them from cell to cell until one player gets his three counters in a row. (I discuss this game in my *Scientific American Book of Mathematical Games and Diversions.*) Moving-counter ticktacktoe is the basis for a number of modern commercial games, such as John Scarne's Teeko and a new game called Touché, in which concealed magnets cause counters to flip over and become opponent pieces.

Standard ticktacktoe can obviously be generalized to larger fields. For example, the old Japanese game of go-moku ("five stones") is essentially five-in-a-row ticktacktoe played on a go board. Another way to generalize the game is to play it on "boards" of three or more dimensions. These variants and others are discussed in my *Wheels, Life* book.

In March, 1977, Frank Harary devised a delightful new way to generalize ticktacktoe. Harary was then a mathematician at the University of Michigan. He is now the Distinguished Professor of Computer Science at New Mexico State University, in Las Cruces. He has been called Mr. Graph Theory because of his tireless, pioneering work in this rapidly growing field that is partly combinatorial and partly topological. Harary is the founder of the *Journal of Combinatorial Theory* and the *Journal of Graph Theory,* and the author of *Graph Theory,* considered the world over to be the definitive textbook on the subject. His papers on graph theory, written alone or in collaboration with others, number more than 500. Harary ticktacktoe, as I originally called his generalization of the game, opens up numerous fascinating areas of recreational mathematics. Acting on his emphatic request, I now call it animal ticktacktoe for reasons we shall see below.

We begin by observing that standard ticktacktoe can be viewed as a two-color geometric-graph game of the type Harary calls an achievement game. Replace the nine cells of the ticktacktoe board with nine points joined by lines, as is shown in Figure 90. The players are each assigned a color, and they take turns coloring points on the graph. The first player to complete a straight line of three points in his color wins. This game is clearly isomorphic with standard ticktacktoe. It is well known to end in a draw if both players make the best possible moves.

Let us now ask: What is the smallest square on which the first player can force a win by coloring a straight (non-diagonal) three-point path? It is easy to show that it is a square of side four. Harary calls this side length the board number b of the game. It is closely related to the Ramsey number of generalized Ramsey graph theory, a number that plays an important part in the Ramsey games. (Ramsey theory is a field in which Harary has made notable contributions. It was in a 1972 survey paper on Ramsey theory that Harary first proposed making a general study of games played on graphs by coloring the graph edges.) Once we have determined the value of b we

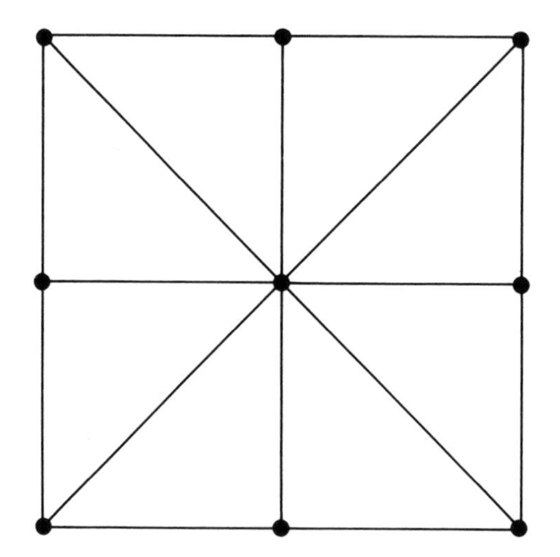

FIGURE 90 Ticktacktoe as a two-coloring game

can ask a second question. In how few moves can the first player win? A little doodling shows that on a board of side four the first player can force a win in only three moves. Harary calls this the move number m of the game.

In ticktacktoe a player wins by taking cells that form a straight, order-3 polyomino that is either edge- or corner-connected. (The corner-connected figure corresponds to taking three cells on a diagonal.) Polyominoes of orders 1 through 5 are depicted in Figures 91 and 92. The polyomino terminology was coined by Solomon W. Golomb, who was the first to make a detailed study of these figures. Harary prefers to follow the usage of a number of early papers on the subject and call them "animals." I shall follow that practice here.

We are now prepared to explain Harary's fortuitous generalization. Choose an animal of any order (number of square cells) and declare its formation to be the objective of a ticktacktoelike game. As in ticktacktoe we shall play not by coloring spots on a graph but by marking cells on square matrixes with noughts and crosses in the usual manner or by coloring cells red and green as one colors edges in a Ramsey graph game. Each player tries to label or color cells that will form the desired animal. The animal will be accepted in any orientation and, if it is asymmetrical, in either of its mirror-image forms.

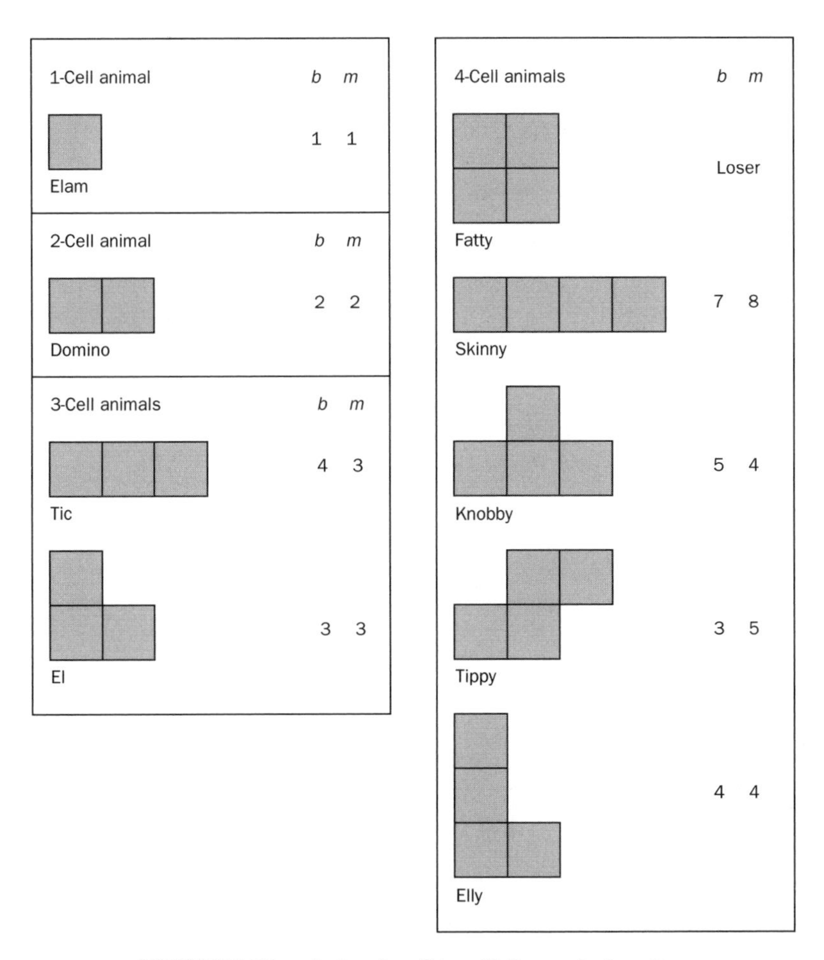

FIGURE 91 Animals of 1 cell through 4 cells

Our first task is to determine the animal's board number, that is, the length of the side of the smallest square on which the first player can, by playing the best possible strategy, force a win. If such a number exists, the animal is called a winner, and it will be a winner on all larger square fields. If there is no board number, the animal is called a loser. If the animal chosen as the objective of a game is a loser, the second player can always force a draw, but he can never force a win. The clever proof of this fact is well known and applies to most ticktacktoelike games. Assume that the second player has a

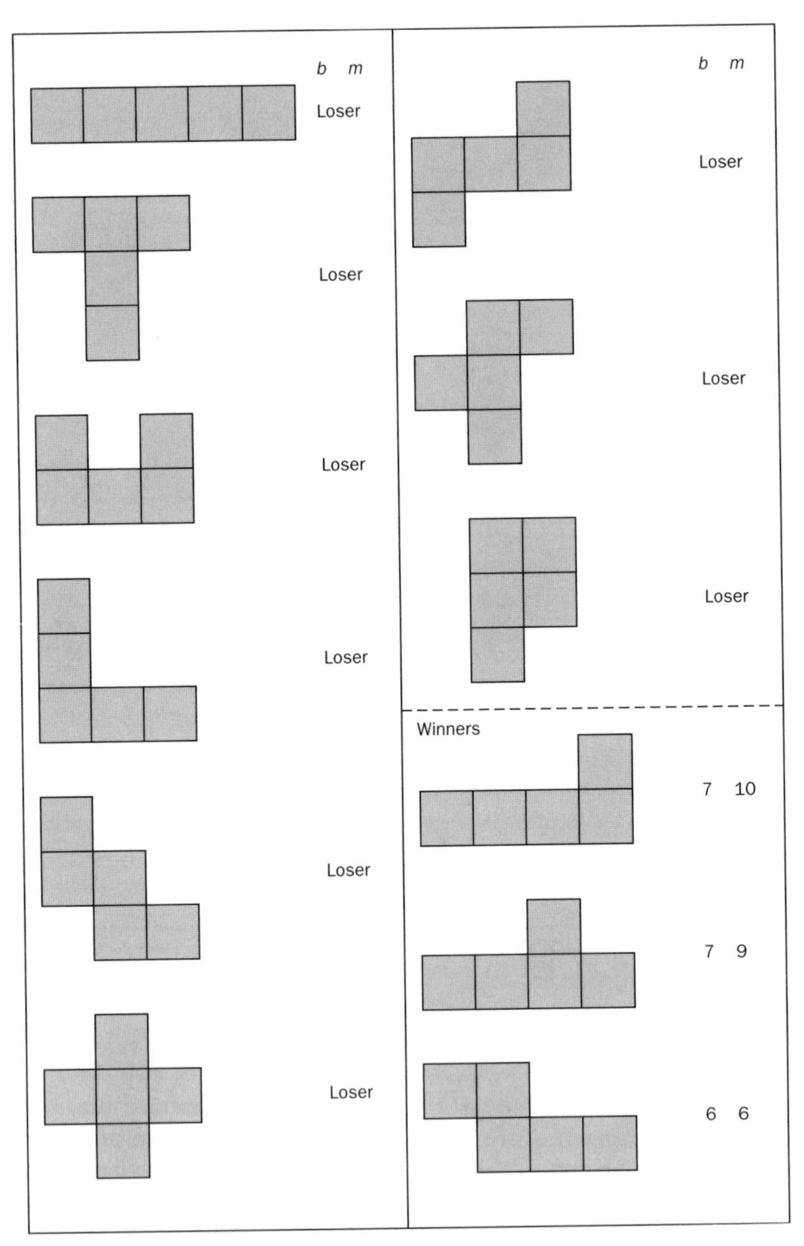

FIGURE 92 The 12 animals of 5 cells

winning strategy. The first player can "steal" the strategy by first making an irrelevant opening move (which can never be a liability) and thereafter playing the winning strategy. This finding contradicts the assumption that the second player has a winning strategy, and so that assumption must be false. Hence the second player can never force a win. If the animal is a winner and b is known, we next seek m, the minimum number of moves in which the game can be won.

For the 1-cell animal (the monomino), which is trivially a winner, b and m are both equal to 1. When, as in this case, m is equal to the number of cells in the animal, Harary calls the game economical, because a player can win it without having to take any cell that is not part of the animal. The game in which the objective is the only 2-cell animal (the domino) is almost as trivial. It is also economical, with b and m both equal to 2. The games played with the two 3-cell animals (the trominoes) are slightly more difficult to analyze, but the reader can easily demonstrate that both are economical: for the L-shaped 3-cell animal b and m are both equal to 3, and for the straight 3-cell animal b equals 4 and m equals 3. This last game is identical with standard ticktacktoe except that corner-connected, or diagonal, rows of three cells are not counted as wins.

It is when we turn to the 4-cell animals (the tetrominoes) that the project really becomes interesting. Harary has given each of the five order-4 animals names, as is shown in Figure 91. Readers may enjoy proving that the b and m numbers given in the illustration are correct. Note that Fatty (the square tetromino) has no such numbers and so is labeled a loser. It was Andreas R. Blass, one of Harary's colleagues at Michigan, who proved that the first player cannot force Fatty on a field of any size, even on the infinite lattice. Blass's result was the first surprise of the investigation into animal ticktacktoe. From this finding it follows at once that any larger animal containing a two-by-two square also is a loser: the second player simply plays to prevent Fatty's formation. More generally, any animal that contains a loser of a lower order is itself a loser. Harary calls a loser that contains no loser of lower order a basic loser. Fatty is the smallest basic loser.

The proof that Fatty is a minimal loser is so simple and elegant that it can be explained quickly. Imagine the infinite plane tiled with dominoes in the manner shown at the top of Figure 93. If Fatty is drawn anywhere on this tiling, it must

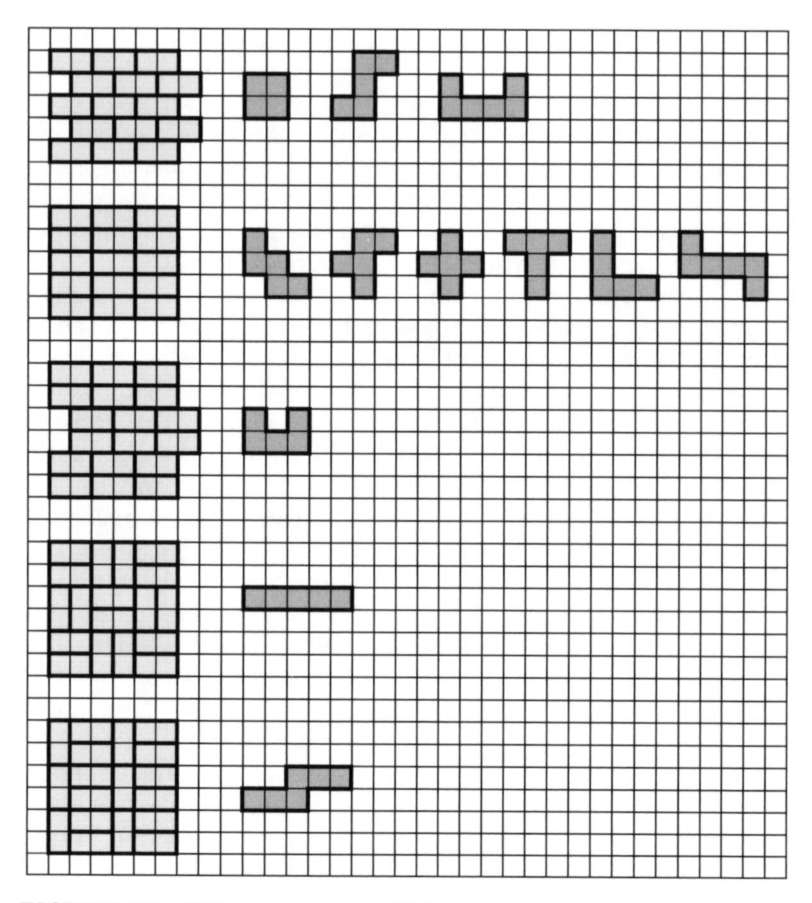

FIGURE 93 Tiling patterns (left) for the 12 basic losers (right)

contain a domino. Hence the second player's strategy is simply to respond to each of his opponent's moves by taking the other cell of the same domino. As a result the first player will never be able to complete a domino, and so he will never be able to complete a Fatty. If an animal is a loser on the infinite board, it is a loser on all finite boards. Therefore Fatty is always a loser regardless of the board size.

Early in 1978 Harary and his colleagues, working with only the top four domino tilings shown in Figure 93 established that all but three of the 12 5-cell animals are losers. Among the nine losers only the one containing Fatty is not a basic loser. Turning to the 35 6-cell animals, all but four contain basic losers of lower order. Of the remaining four possible winners three can be proved losers with one of the five tilings shown in the illustration. The animals that can be proved basic losers with each tiling pattern are shown alongside the pattern. In every case the proof is the same: it is impossible to draw the loser on the associated tiling pattern (which is assumed to be infinite) without including a domino; therefore the second player can always prevent the first player from forming the animal by following the strategy already described for blocking Fatty. There are a total of 12 basic losers of order six or lower.

It is worth noting how the tiling proof that the straight animal of five cells is a loser (another proof that was first found by Blass) bears on the game of go-moku. If the game is limited to an objective of five adjacent cells in a horizontal or vertical line (eliminating wins by diagonal lines), the second player can always force a draw. When diagonal wins are allowed, the game is believed to be a first-player win, although as far as I know that has not yet been proved even for fields larger than the go board.

The only 6-cell animal that may be a winner is the one that I named Snaky:

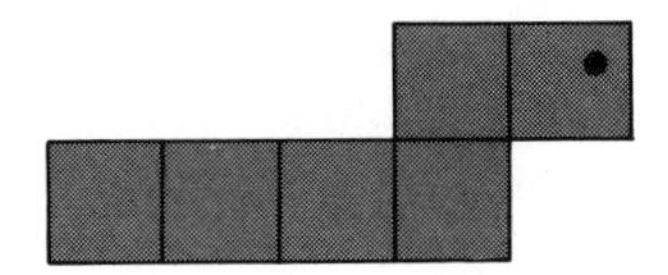

Although they have not yet been able to prove this animal is a winner, they conjecture its board number, if any, is no larger than 15 and its move number is no larger than 13. This assertion is the outstanding unsolved problem in animal ticktacktoe theory. Perhaps a reader can prove Snaky is a loser or conversely show how the first player can force the animal on a square field and determine its board and move numbers.

All the 107 order-7 animals are known to be losers because each contains a basic loser. Therefore since every higher-order animal must contain an order-7 animal, it can be said with confidence that there are no winners beyond order 6. If Snaky is a winner, as Harary and his former doctoral student Geoffrey Exoo conjecture, there are, by coincidence, exactly a dozen winners—half of them economical—and a dozen basic losers.

Any 4- or 5-cell animal can be the basis of a pleasant pencil-and-paper game or a board game. If both players know the full analysis, then depending on the animal chosen either the first player will win or the second player will force a draw. As in ticktacktoe between inexpert players, however, if this knowledge is lacking, the game can be entertaining. If the animal chosen as the objective of the game is a winner, the game is best played on a board of side b or $b-1$. (Remember that a square of side $b-1$ is the largest board on which the first player cannot force a win.)

All the variations and generalizations of animal ticktacktoe that have been considered so far are, as Harary once put it, "Ramseyish." For example, one can play the misère, or reverse, form of any game—in Harary's terminology an avoidance game—in which a player wins by forcing his opponent to color the chosen animal.

Avoidance games are unusually difficult to analyze. The second player trivially wins if the animal to be avoided is the monomino. If the domino is to be avoided, the second player obviously wins on the 2×2 square, and almost as obviously on the 2×3 rectangle.

On a square board of any size the first player can be forced to complete the L-shaped 3-cell animal. Obviously the length of the square's side must be at least 3 for the game to be meaningful. If the length of the side is odd, the second player will win if he follows each of his opponent's moves by taking the cell symmetrically opposite the move with respect to the center of the board. If the first player avoids taking the center, he will be forced to take it on his last move and so will lose. If he takes it earlier in the game without losing, the second player should follow with any safe move. If the first player then takes the cell that is symmetrically opposite the second player's move with respect to the center, the second player should again make a harmless move, and so on; otherwise he should revert to his former strategy. If the length of the square's side is even, this type of symmetrical play leads to a draw, but the second player can still win by applying more complicated tactics.

On square boards the straight 3-cell animal cannot be forced on the first player. The proof of this fact is a bit difficult, even for the three-by-three square, but as a result no larger animal containing the straight 3-cell species can be forced on any square board. (The situation is analogous to that of basic losers in animal-achievement games.) Hence among the 4-cell animals only Fatty and Tippy remain as possible nondraws. Fatty can be shown to be a draw on any square board, but Tippy can be forced on the first player on all square boards of odd side. The complete analysis of all animal-avoidance games is still in the early stages and appears to present difficult problems.

Harary has proposed many other nontrivial variants of the basic animal games. For example, the objective of a game can be two or more different animals. In this case the first player can try to form one animal and the second player the other, or both players can try to form either one. In addition, achievement and avoidance can be combined in the same game, and nonrectangular boards can be used. It is possible to include three or more players in any game, but this twist introduces coalition play and leads to enormous complexities. The rules can also be revised to accept corner-connected animals or animals that are both edge- and corner-connected. At the limit, of course, one could make any pattern whatsoever the objective of a ticktacktoelike game, but such broad generalizations usually lead to games that are too complicated to be interesting.

Another way of generalizing these games is to play them with polyiamonds (identical edge-joined equilateral triangles) or polyhexes (identical edge-joined regular hexagons) respectively on a regular triangular field or a regular hexagonal field. One could also investigate games played with these animals on less regular fields. An initial investigation of triangular forms, by Harary and Heiko Harborth, is listed in the bibliography.

The games played with square animals can obviously be extended to boards of three or more dimensions. For example, the 3-space analogue of the polyomino is the polycube: n unit cubes joined along faces. Given a polycube, one could seek b and m numbers based on the smallest cubical lattice within which the first player can force a win and try to find all the polycubes that are basic losers. This generalization is almost totally unexplored, but see the bibliography for a paper on the topic by Harary and Michael Weisbach.

As I have mentioned, Blass, now at Pennsylvania University, is one of Harary's main collaborators. The others include Exoo, A. Kabell and Heiko Harborth, who is investigating games with the triangular and hexagonal cousins of the square animals. Harary is still planning a book on achievement and avoidance games in which all these generalizations of ticktacktoe and many other closely related games will be explored, and he is also persuading his current computer science students to develop computer programs for playing these games both offensively and defensively. This is the area of AI (artificial intelligence) known as game-playing programs.

ADDENDUM

In giving the proof that a second player cannot have the win in most ticktacktoe-like games, I said that if the first player always wins on a board of a certain size, he also wins on any larger board. This is true of the square boards with which Harary was concerned, but is not necessarily true when such games are played on arbitrary graphs. A. K. Austin and C. J. Knight, mathematicians at the University of Sheffield, in England, sent the following counterexample.

Consider the graph at the left of Figure 94, on which three-in-a-row wins. The first player wins by taking *A*. The second player has a choice of taking a point in either the small or the large triangle. Whichever he chooses, the first player takes a corner point in the other triangle. The opponent must block the threatened win, then a play in the remaining corner of the same triangle forces a win.

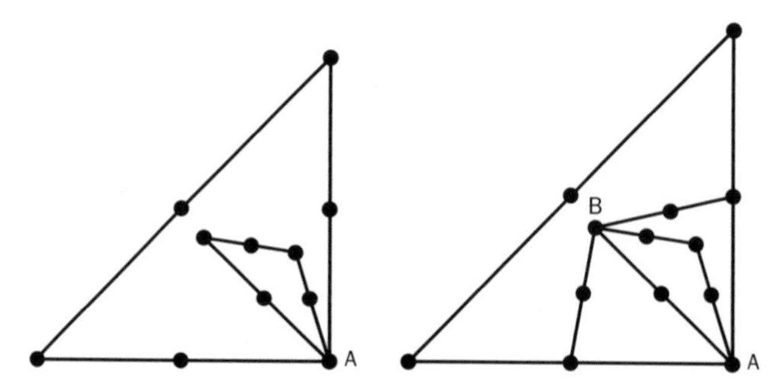

FIGURE 94 First player wins on graph at left, but second player can force a draw on enlarged graph at right

Now enlarge the "board" by adding two points as shown on the right in Figure 94. The second player can draw by playing at *B*. If the first player does not start with *A*, the second player draws by taking *A*.

Achievement and avoidance games played on graphs obviously open up endless possibilities that will be explored in Harary's forthcoming book.

BIBLIOGRAPHY

Polyominoes. Solomon W. Golomb. Scribner's, 1965.

Graph Theory. Frank Harary. Addison-Wesley, 1969.

"Extremal Animals." Frank Harary and Heiko Harborth, in *Journal of Combinatorics, Information, and System Sciences*, 1, 1976, pages 1–8.

"Polycube Achievement Games." Frank Harary and Michael Weisbach, in *Journal of Recreational Mathematics*, 15, 1982–83, pages 241–246.

"Achieving the Skinny Animal." Frank Harary, in *Eureka*, No. 42, Summer, 1982, pages 8–14; Errata, No. 43, Easter, 1983, pages 5–6.

"Ticktacktoe Games." Martin Gardner, in *Wheels, Life, and Other Mathematical Amusements*, Chapter 5. W. H. Freeman, 1983.

"Achievement and Avoidance Games with Triangular Animals." Frank Harary and Heiko Harborth, in *Journal of Recreational Mathematics*, 18, 1985–86, pages 110–115.

Achievement and Avoidance Games. Frank Harary. Book in preparation.

14

Psychic Wonders and Probability

Conjuring is the art of entertaining people by doing things that seem to violate natural laws. A woman levitates. An elephant vanishes. Doves materialize. Spoons bend when stroked with a finger. And so on.

A small branch of magic, much in the news these days because of public enthusiasm for self-styled psychics, is concerned not with doing the impossible but only with doing something extremely improbable. Feats of this kind usually call on things that serve as randomizers in games of chance, such as a deck of cards or a pair of dice, so that there is an understandable overlap between the methods followed by psychic charlatans and those followed by gambling hustlers and honest charlatans (magicians).

A full account of magic and probability would involve a multivolume discussion of ways of controlling dice, cards, flipped coins, roulette wheels, bingo and other games of chance, and of "gaffing" carnival games. In all these areas there are cheating stratagems of incredible ingenuity. As an example, a vertically spinning carnival wheel can be gaffed with the aid of a loose floorboard. One end of the board presses upward against the bottom of a vertical board that runs behind the wheel and that has a a hole in it through which the rotating axle of the wheel extends. The carnival operator stands at the side of his booth opposite the wheel. By shifting his weight imperceptibly from one leg to the other, he can manipulate the loose board as a long lever and can push the vertical board up to put friction on the axle. The beauty of this particular gaff is that no one can prove that it is intentional even if the entire booth is torn apart.

In this chapter I shall skim briefly over a few of the hundreds of ways in which fake psychics can boost their odds when they work with playing cards or ESP cards. Today any sincere parapsychologist would make certain that a subject does not see or handle the cards, but in the early days of parapsychology, when the most sensational results were obtained, such controls were often not imposed. Even today in informal testing and in public demonstrations by dishonest psychics the ESP cards are sometimes in full view.

What can a clever psychic do to raise his scores above chance levels? One method is to secretly mark some of the cards in the course of handling them during preliminary tests. There are many ways of doing it. Card hustlers often resort to what in the trade is known as daub, a waxy substance that can be put on the back of a coat button for easy access to a fingertip. It leaves on the border of the card an extremely faint smudge indistinguishable from the smudges cards acquire in ordinary handling. Another method is to put a tiny nick on the edge of the card with a fingernail. A third method is to scrape a thumbnail about an eighth of an inch along the edge to produce a whitish line. It is only necessary to mark a few cards in order to raise a score to significant levels.

Sometimes it is not even necessary to mark cards to recognize them from the back. After playing cards or ESP cards have been used they tend to acquire all kinds of minute imperfections: a dirt mark, a slightly bent corner and so on. That is why in professional card games new decks are constantly being produced. Suppose a sharp-eyed poker player notices that the king of hearts has a tiny speck of dirt at one corner.

If in a game of stud poker he sees the speck on an opponent's hole card, is he cheating or just being alert?

It is not generally known, even by magicians, that the official ESP cards now in use (authorized by J. B. Rhine) have what card magicians call "one-way backs." This means that if you examine the backs carefully, you will find they are not the same when the card is rotated 180 degrees. For example, the upper right-hand corner of the back of an ESP card either has a star there or it does not. In the course of a trial run there are many ways a psychic can set a 25-card deck so that all its cards are "one way." For example, he will try to guess only the cards that are turned one way and will not try on all the others. The unguessed cards are dealt to a separate pile. After the test (on which he is likely to score at the level of random chance) one pile is turned around before the pack is reassembled. The cards are now all one way, and the psychic is ready to perform miracles.

The literature of card magic is filled with clever tricks based on the one-way principle. A psychic may spread the cards on a table, turn his back and ask someone near the left end of the table to draw a card. That person is then told to hand the card to someone near the right end of the table to verify the symbol. The second person then returns the card to the spread and gives the deck a shuffle. This maneuver nearly always reverses the card. (There are many other subtle procedures for causing one or more chosen cards to become reversed in a one-way deck.) The cards are then dealt in a row. The psychic turns around, and moving his hand slowly down the row to "feel the vibes," he easily locates the chosen card. The card is turned face up by rotating it end for end and is then replaced in the row by turning it face down from side to side, so that the cards are one way again.

It is not necessary to have the ESP deck all one way to perform similar feats. Suppose the deck is random with respect to its back patterns and that you are the psychic. Have five cards dealt in a row. Simply memorize the pattern as a binary number, say 11010. Follow the above strategy for having a card selected and returned to the row while your back is turned. You can find the card easily, and you can memorize the new binary pattern and repeat the performance as often as you like.

A standard deck of 25 ESP cards has five cards for each symbol: a star, a cross, a circle, wavy lines and a square. It is only necessary to have the 10 cards for two symbols one way and the 15 cards for the other three symbols the other way to make an impressive score in guessing each symbol as someone deals the shuffled pack face down. For an unprepared deck the expected score is five hits, assuming there is no feedback information about the cards until the dealing is completed. With two symbols set one way you have a chance of 1/2 of guessing correctly between the two. This is an expectation of 10/2, or five, hits. You have a 1/3 chance of guessing correctly on the other three symbols, which is an expectation of 15/3, or again five, hits. The total expectation is five plus five, or 10, hits for the run of 25 cards. If the run is repeated four times, for the standard four-run test you can expect 40 hits out of 100. Parapsychologists regard a score of 30 as indicating excellent ESP, and so a score of 40 is sensational. It is much more impressive than a perfect score, which would strongly suggest cheating. Note the curious fact that if only one symbol is marked, your expectation is still 40 hits in a run of 100. That, however, requires always guessing one symbol correctly, which would be noticed on score sheets and give the game away.

If the cards are not visible to the psychic in a test, a common method of cheating is to rely on what magicians call a "stooge": someone who is watching behind a screen and sending secret signals to the psychic by any one of scores of little-known techniques. (One of the hardest to detect involves the electronic sending of pulses that the psychic receives by means of a tiny device inserted in the anus. Few examinations of psychics take this possibility into account.)

The use of secret stooges is common in the history of psychic phenomena. Almost all "mind reading" animals, which in the past included famous vaudeville horses, dogs and even pigs (see Figure 95) rely on secret sound signals that can be picked up easily by species with large ears and a sense of hearing better than man's. For example, the trainer has a playing card selected and learns the card by standard magic methods. The animal then "checks" each of 52 cards spread in a large circle on the floor or displayed on a long easel. When it comes to the right card it paws it, noses it or picks it up in its mouth. The card is of course cued by a signal only the animal can hear. It may be almost anything, from a faint sniffing sound to the clicking of one fingernail against another by a hand held in a pocket or behind the back. The signaler need

FIGURE 95 A mind-reading pig is advertised in this poster in the collection of Milbourne Christopher

not be the person on stage; it can be someone sitting in the first row.

The trained animal can also pick alphabet cards to spell a word or numbered cards that display the solution to a mathematical problem called out by a spectator. Many old books go into great detail on how to train animals for such acts. One of the best is *Haney's Art of Training Animals* (Jesse Haney & Co., Publisher, 1869), which recommends fingernail signaling. Another is *The Expositor: Or Many Mysteries Unravelled,* by William Frederick Pinchbeck (privately published, 1805). It has an excellent section on training mind-reading pigs by the sniffing-sound method.

When J. B. Rhine was a young man he was completely taken in by Lady Wonder, a mind-reading horse in Virginia whose psi abilities he and his wife enthusiastically described in *The Journal of Abnormal and Social Psychology* (Vol. 23, 1929, pp. 449–466). A more sophisticated Rhine later repeated his tests with Lady Wonder and found that her trainer was signaling. Until his death Rhine contended that Lady Wonder did have genuine extrasensory perception but that after she had lost it her owner began signaling to her in secret. (For details see my *Fads and Fallacies in the Name of Science,* a Dover paperback, pages 351–352.)

It is usually impossible to know whether or not stooges were present by reading a parapsychologist's official report of an experiment. If you go over the papers by Harold E. Puthoff and Russell Targ that recount their clairvoyance tests of the Israeli magician Uri Geller at the time he was visiting the Stanford Research Institute, you will find no hint that Geller's best friend, Shipi Strang, was always present during the tests. When this fact came out, Puthoff and Targ agreed that he was there, but they wrote that Strang was carefully "excluded from the target area." By this they meant that he was not in the room where the randomly selected target pictures were being "sent" to Uri. But Strang, it turns out, was unattended and moving around freely, and there are many ways he could have learned of the targets and signaled them. The contrast between the elliptical, sanitized reports of the experiments and the chaotic conditions that actually prevailed, as described in John Wilhelm's *The Search for Superman* (Pocket Books, 1976), and James Randi's *The Magic of Uri Geller* (Ballantine Books, 1976), is startling. That Strang often stooged for Geller is well established, having been described by Strang's sister in an interview she gave an Israeli reporter and also by Geller's former manager, Yasha Katz, in an Italian television interview in 1979.

If a psychic is guessing ESP cards that are concealed from him, the cards do not have to be marked or turned one way for a stooge to send valuable information. It is only necessary for the stooge to signal the symbol after the card is turned to enable the psychic to score well above chance. With complete information about each card after the guess has been made a psychic who adopts his best strategy can raise his expected score on 25 cards from five to 8.65. (See Ronald C. Read's article listed in the bibliography.)

If the feedback is no more than whether the psychic made a hit or a miss, it allows a strategy capable of producing an expected score of 6.63 hits. (See the paper by Persi Diaconis). In many classic tests of ESP it is impossible to tell from published reports whether a subject was given any kind of feedback or whether friends of the subject were present who could have sent feedback by secret signaling.

In most card games, feedback from dealt cards gives valuable information to a skilled player. That this is the case in stud poker and bridge is obvious. In the casino game of blackjack, or twenty-one, a player capable of memorizing the values of dealt cards can actually win consistently by betting high when the odds favor him and low when they do not. Such players are called counters, and in recent years they have done so well that most casinos now refuse to let known counters play the game. It hardly seems fair, because someone who is counting is not in any reasonable sense of the word cheating.

Intuition can go wildly astray in evaluating feedback from dealt cards. I know of no more startling example than a card-betting game discovered by Robert Connelly, a mathematician at Cornell University who made news in 1978 by disproving a famous conjecture about polyhedrons. If a polyhedron has rigid faces but is hinged along all its natural edges, can it be "flexed," that is, made to alter its shape? It had been known since 1813 that the answer is no if the polyhedron is convex, and it had been conjectured that the same was true of all nonconvex polyhedrons with nonintersecting faces. Connelly found a counterexample with 18 triangular faces.

Connelly calls his game Say Red. The banker shuffles a standard deck of 52 cards and slowly deals them face up. The dealt cards are left in full view where they can be inspected at any time by the player. Whenever the player wants, he may say "Red." If the next card is red, he wins the game, otherwise he loses. He must call red before the deal ends, even if he waits to call on the last card. What odds should the banker give to make it a fair game, assuming that the player adopts his best strategy on the basis of feedback from the dealt cards? The player must announce the size of his bet before each game begins.

Another well-kept secret of psychic charlatans for increasing their probability of success is the use of what Diaconis, in the paper mentioned above, calls multiple end points. Magicians know them more informally as "outs." The basic idea is for the psychic not to specify in advance exactly what he plans to do and then let the outcome depend on what happens. As Diaconis puts it, the chance for some kind of coincidence is very much better than the chance for a coincidence specified in advance. Diaconis describes the work of a psychic known in the literature as "B. D.," whose playing-card feats were favorably written about in three papers in Rhine's *Journal of Parapsychology*. Diaconis, a former magician of great skill with cards, was present as an observer during one of B. D.'s demonstrations at Harvard University. Diaconis recognized at once that B. D.'s main secret was his reliance on outs.

Although I cannot go into detail here about how magicians who pretend to be psychics make extensive use of outs, I can illustrate the technique with an anecdote followed by two remarkable "precognition" tricks with which readers can amaze their friends. Paul Curry, a New York amateur magician, likes to remember the time a magician friend had been asked to do some card tricks at a party. He noticed that the deck handed to him included a card, say the eight of clubs, that was badly torn. Since such a card would have interfered with his manipulations, he surreptitiously removed it and put it in his pocket. During his performance he asked someone to name a card. A woman called out: "The eight of clubs." Without thinking he responded: "I'm sorry, madam, but you'll have to call another card. You see, before I started performing I noticed that the eight of clubs was torn and so I—" He broke off and almost fainted when he realized he had missed a chance to work a miracle: to make the named card vanish from the deck and reappear in his pocket, albeit damaged in transit.

It is hard to believe, but magicians have actually devised tricks with a stacked deck (a deck in prearranged order) with 52 different outs depending on the card named! The card can be shown on the top, on the bottom or as the only reversed card in the deck. One can spell or count to the card in various ways, discover a duplicate of the card tacked to the ceiling, take from the breast pocket a handkerchief and open it to show on it a large picture of the card, have a spectator look under the cushion of his seat and find a duplicate of the card, show that it is the only card missing from the deck, toss the deck against a window blind and have the blind shoot up to reveal the card pasted outside the window, and so on.

The following clever card trick, calling for six cards and six outs, has several forms. The version given here was devised by Tom Ransom of Toronto. The six cards are arranged in a row. You can tell your audience honestly that the cards bear values from 1 through 6. All the cards except one have a blue back. The two of spades has a red back and is face up. All the cards are spades except the five of hearts, which is of course face down. Write on a piece of paper, "You will choose the red card," and put the prediction face down.

Ask for any number from 1 through 6. If you like, you can hand someone an "invisible die" and ask him to pretend to roll it and tell you what number comes up. Here are the six outs for each of the six numbers:

1. From your left count to the first card. Turn it over to show its red face. Turn over the other face-down cards to show that all their faces are black.

2. Turn over the deuce to show its red back. Reverse the other two face-up cards to show that all the other cards have blue backs.

3. Ask someone to count to the third card from his left. Finish as above.

4. Count to the fourth card from your left. Finish as above.

5. Turn over all the face-down cards to disclose that the five-spot card is the only one with a red face.

6. Ask the other person to count from his left to the sixth card. Finish as above.

No matter what number is chosen your prediction will be accurate. You cannot, of course, repeat the trick for the same audience.

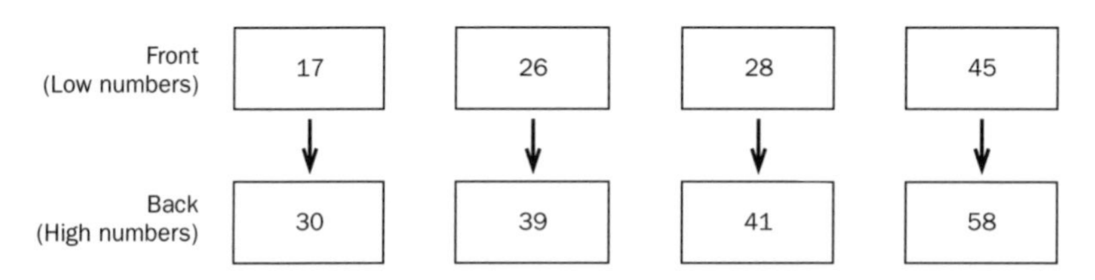

FIGURE 96 Shigeo Futigawa's sum prediction trick

Another ingenious trick involving outs is currently on sale in Japanese magic shops. It is the invention of Shigeo Futigawa, a Tokyo mathematics teacher and amateur magician.

You will need four identical cards that are blank on both sides. On one card print the number 17 and on the back of it print the number 30. On the other three cards print the number pairs 26/39, 28/41 and 45/58. In the first row of Figure 96 are four cards that have been put down with their low number uppermost. You must memorize these four low numbers so that you can identify them quickly, or if you prefer, you can pencil a dot somewhere near each of them to distinguish them from the high numbers on the back of the cards.

Hand the four cards to someone with the request that he mix them thoroughly by turning the cards over any way he likes and that he then place the cards on the table so that the four visible numbers have been randomly selected. Before he does so you write a predicted total on a piece of paper and put it face down to one side.

When the randomized cards are on the table, there are just three possibilities. For each you follow a different procedure:

1. The cards may show two low numbers and two high ones. In the long run this will happen three out of eight times. Have the four numbers added. The sum will be 142. That is what you wrote on the paper. Let someone check your prediction.

2. All four cards may be high or all four low. This happens one out of eight times. Stand up, turn your back and ask someone to randomize the numbers a bit more by turning over any pair of cards. That, of course, alters them to two high and two low, and so you finish as before.

3. The cards may show either three high and one low or three low and one high. This, perhaps surprisingly, is the commonest pattern, occurring with a probability of exactly 1/2. When it happens, pick up the single card (either the sole low card or the sole high card) and as you turn it over say: "Please notice that the numbers on opposite sides of each card are not the same. If you had placed this card down the other way, the sum of the four visible numbers would have been entirely different." You now stand up, turn your back and ask someone to reverse any two cards to randomize them further. Because that yields an even split of high and low numbers, you can finish as you did in the first case.

In brief, the procedure you adopt always yields a sum of 142, so you can't miss. Note that the difference between each pair of numbers on a card is 13. I leave it to the reader to prove algebraically that the trick must work. The difference can be any number. The predicted total will be the sum of the four low numbers plus twice the difference. This makes it easy to prepare cards with numbers other than the eight I have given here.

Is it possible to prepare four cards with eight different numbers so that the trick can be done exactly the same way except that the four selected numbers are multiplied instead of added? You must be able to predict the product the same way you predicted the sum, and by following the same three outs. The answer is yes, and Futigawa has designed such a set of cards as a variant of his earlier trick. I shall describe this second set in the answer section and explain the secret of its construction.

ANSWERS

The first problem concerned Robert Connelly's card betting game Say Red. In the game a banker deals cards face up from a standard deck, and at some point (before the last card is dealt) the player must say "Red." If the next card is red, he wins the game; otherwise he loses. At all times the player may inspect the cards already dealt. From this information can he devise a strategy that will in the long run raise his probability of winning above 1/2?

Surprisingly, he cannot. There are many formal proofs of this fact, but perhaps the best way to "see" why it is true is to analyze a deck consisting of two black and two red cards. Labeling the black cards B and the red cards R, the six equally likely deals are as follows.

RRBB
RBRB
RBBR
BBRR
BRBR
BRRB

Glancing down the column, it is obvious that the probability the player will win is 1/2 regardless of whether he calls before the first, the second, the third or the last card has been dealt. On the other hand, suppose two black cards are dealt. One could argue that the third must be red and therefore the player cannot lose. That is true, but in the long run it happens only once in six deals. On the other five deals the probability that the player will win is $5/6 \times 2/5$, or 1/3, and so the overall probability of winning on the third card is $1/6 + 1/3$, or 1/2.

Now try a different tack. Suppose the player's strategy is to call red after the first card is dealt if that card is black, and otherwise to call red before the last card is dealt. If the first card is black, the player does have a 2/3 chance of winning on the next card, but this is counterbalanced by the fact that if the first card is not black, he has a 2/3 chance of losing on the last card.

This reasoning generalizes to decks with b black cards and r red ones. The probability that a player will win by any strategy whatsoever, Connelly points out, is always $r/(b+r)$. Therefore knowing which of the cards in a deck have been dealt gives a player no advantage at all, regardless of the ratio of the black cards to the red cards or the size of the deck. No strategy is better than always calling red on the first card, or on the last card, or indeed on any card. (As I stated, the player must specify his bet before each deal; otherwise he could come out ahead by, as in a system for playing blackjack, betting high when the odds favor him and low when they do not.)

The second problem was to design a set of four cards for a magic trick similar to the one described except that the four numbers apparently picked at random must always have the same product rather than the same sum. The secret is to choose pairs of numbers to be placed on opposite sides of the cards that all have the same ratio. For example, on the cards prepared by Shigeo Futigawa, the Tokyo magician who invented both variants of the trick, the number pairs are 26/34, 39/51, 52/68 and 65/85. In each case the ratio is 13 to 17.

The trick is handled almost exactly the way the one I described earlier is, except that the four selected numbers are multiplied instead of added. The predicted product in this case is 5,860,920. Note that this number is equal to $13\times 13\times 17\times 17\times 2\times 3\times 4\times 5$. As with the addition version, I leave it to readers to prove algebraically that the trick cannot fail.

ADDENDUM

Several versions of the red card prediction have appeared in magic literature in which more than six cards are in the row. In my opinion they all suffer from the use of spelling the names of numbers instead of counting. Everybody should realize that one, two, six and ten have three letters; four, five and nine have four letters; and three, seven, and eight have five letters. This weakens the use of spelling for any of the outs. An example using such spellings is a ten-card version by "Corinda," in his book *13 Steps to Mentalism.* Magicians have also devised ways of apparently showing the six cards, in the version I gave, as the ace through six of spades, and all with blue backs. The later appearance of a red deuce or a red-backed five then comes as a magical surprise.

R. L. Dreyfus suggested that in Futigawa's prediction the cards could have low even numbers on one side and high odd numbers on the other. The advantage would be that it is easier to distinguish even from odd than high from low.

Persi Diaconis pointed out that the sniff signal, used in calculating animal acts, is also commonly used by two collaborating hustlers in poker, bridge, and other card games. It alerts the partner for a secret gesture that will indicate what is in the signaler's hand.

BIBLIOGRAPHY

"Card Guessing with Information—a Problem in Probability." Ronald C. Read, in *American Mathematical Monthly,* 69, 1962, pages 506–511.

"Magic and Paraphysics." Martin Gardner, in *Science: Good, Bad, and Bogus,* Chapter 8. Prometheus Books, 1981.

"Statistical Problems in ESP Research." Persi Diaconis, in *Science,* 201, 1978, pages 131–136.

"A Flexible Sphere." Robert Connelly, in *The Mathematical Intelligencer,* 1, 1978, pages 130–131.

How Not to Test a Psychic. Martin Gardner. Prometheus Books, 1989.

15

Mathematical Chess Problems

"It is a beautiful, complex and sterile art related to the ordinary form of the game only insofar as, say, the properties of a sphere are made use of both by a juggler in weaving a new act and by a tennis player in winning a tournament. Most chess players . . . are only mildly interested in these highly specialized, fanciful, stylish riddles."

—VLADIMIR NABOKOV, *Speak, Memory*

A chapter in my *Wheels, Life, and Other Mathematical Amusements* is devoted to curious chess problems. The topic of this chapter is again chess, in the form of some unusual chess puzzles I have not dealt with before. Here "chess puzzle" does not refer to the conventional type of chess problem that asks how a player can checkmate in a certain number of moves. The tasks to be considered are of a more mathematical nature, so that to work on them readers need only know the rules of the game.

I shall begin with a remarkable discovery made by David A. Klarner, a mathematician now at the University of Nebraska, in Lincoln. In the early 1960's pure-number theorists were working on a problem concerning any set of consecutive integers from 1 through $2n$, where n is a positive integer. Such a set can always be divided into two equal subsets, one made up of the consecutive integers 1 through n and the other made up of the consecutive integers $n+1$ through $2n$. The problem asks whether for different values of n it is possible to pair each number in one subset with a number in the other so that the $2n$ sums and absolute (unsigned) differences of the numbers in each pair are all distinct. For example, if $n=1$, then the original set is $\{1,2\}$, and its subsets are $\{1\}$ and $\{2\}$. In this case the only pairing provides a trivial solution, because the sum $1+2$, or 3, is distinct from the absolute value of the difference $1-2$, or 1.

Klarner's clever discovery was that this problem is isomorphic with a variation of the old chess problem of how to put n chess queens on a board of side n in such a way that no two queens attack each other. He modified the task by placing an extra row at the top of the chessboard. He calls the row a reflection strip, because, as is shown with a dashed line in Figure 97, one queen can attack another by reflection in the strip. In other words, a queen is allowed to move diagonally into the strip and out again on the opposite diagonal to attack another queen. In Klarner's paper "The Problem of Reflecting Queens" he proved that the number-theory problem for a given n has a solution if and only if it is possible to place n queens on an n-by-n board with a reflection strip in such a way that no two queens attack each other either directly or by reflection.

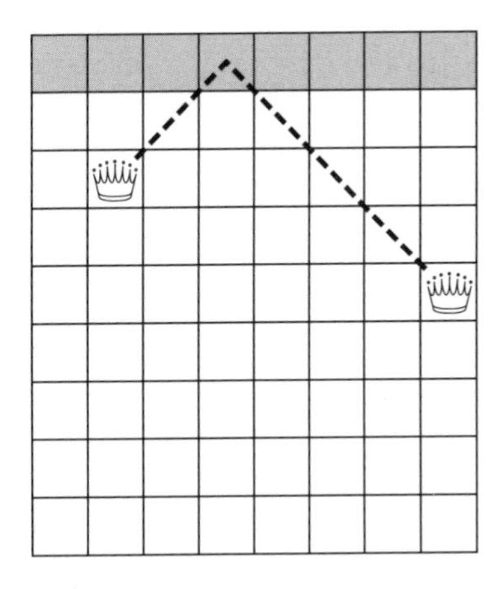

FIGURE 97 Two queens attacking by reflection

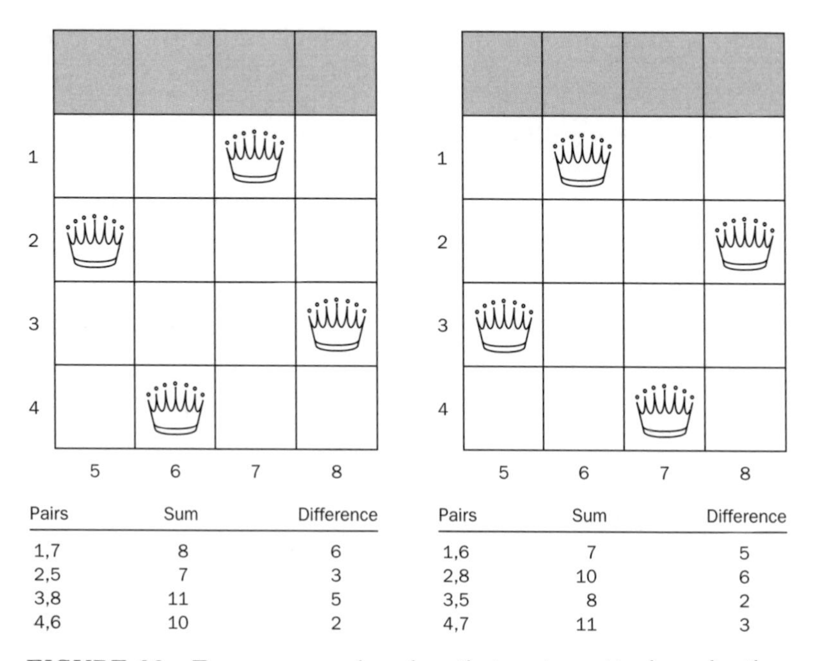

Pairs	Sum	Difference
1,7	8	6
2,5	7	3
3,8	11	5
4,6	10	2

Pairs	Sum	Difference
1,6	7	5
2,8	10	6
3,5	8	2
4,7	11	3

FIGURE 98 Four queens placed so that no two attack each other directly or by reflection

Figure 98 shows how easily a solution of this variation on the queen-placement problem (in this instance for the case $n=4$) can be transformed into a solution of the number problem. Beginning at the upper left on the board, number the rows consecutively from top to bottom and then continue the sequence from left to right below the columns, as is shown in the illustration. Note that the two solutions shown (the only ones for this case) are left-to-right mirror images. Now make a list of the row and column numbers, or coordinates, for each queen, as is shown below the boards in the illustration. Taking the sums and differences of the eight pairs of numbers obtained in this way, it is easy to verify that they provide a solution to the number-theory problem for the case $n=4$.

It is not hard to understand why the two problems are isomorphic. It is obvious that the set of numbers listed around the outside of the chessboard is equal to the set of numbers 1 through $2n$ and that the row numbers for the queens in a solution form one subset of the set and the column numbers form another subset. If two queens did attack each other in the same row or column, then a number would be duplicated in one of the subsets. Since this cannot happen, the subsets must respectively be made up of the numbers 1 through n and the numbers $n+1$ through $2n$. Moreover, if two queens attacked on a diagonal slanting up and to the right, one of the sums would be repeated. If two queens attacked on a diagonal slanting down and to the right, an absolute difference would be repeated. And if two queens attacked by reflection in the strip, a sum and an absolute difference would be equal. Since no two queens may attack each other directly or by reflection, however, the sums and absolute differences of the pairs of numbers derived from the chess solution must all be distinct, and so they provide a solution to the number-theory problem.

Solutions to the classic queens problem (with no reflection strip) on boards of up through order eight have long been known and are discussed in Chapter 16 of my book *The Unexpected Hanging and Other Mathematical Diversions*. There are no solutions for orders two and three and, ignoring rotations and reflections, only one "basic" solution for order four, two basic solutions for order five, one for order six, six for order seven and twelve for order eight. To derive all the solutions for the corresponding number problem from one of these sets of solutions it is necessary to consider the four rotations of each basic pattern and their left-to-right mirror images, ruling out all cases in which queens attack by reflection in a strip added to the board. The remaining cases provide a complete set of solutions to the number problem.

For the case $n=4$ the pattern of the four queens that is the single solution to the basic problem has rotational symmetry, so that there is no need to test rotations. For $n=5$ the five queens can be arranged in the two essentially different ways shown in Figure 99. The pattern at the left contains a reflection attack (shown by the dashed line) and has rotational symmetry. Therefore it cannot in any of its rotations or reflections generate a solution to the number problem. The second pattern generates one solution in the orientation shown and another when it is rotated clockwise 90 degrees. The mirror images of these two patterns give two other solutions, making four in all.

The single pattern of nonattacking queens in the case $n=6$ has reflection attacks in all rotations (and in their mirror images), so that there is no solution to the corresponding number problem. Therefore the following playing-card puzzle can-

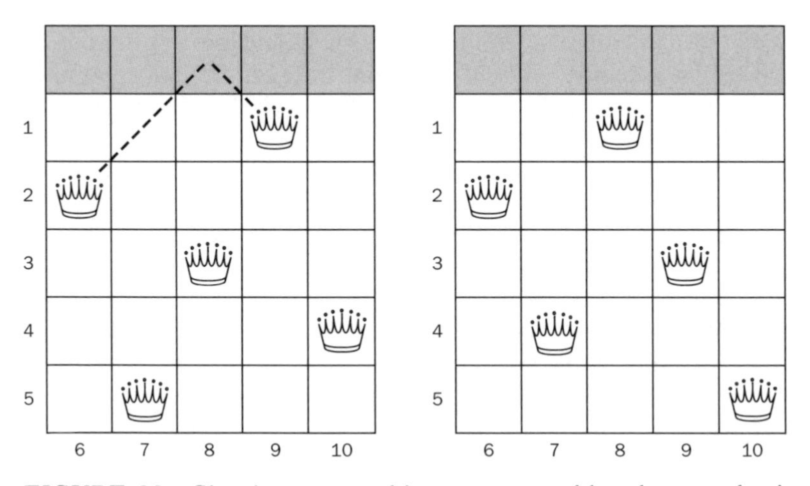

FIGURE 99 Classic non-attacking queens problem has two basic solutions for the case $n=5$

not be solved. Place in a row cards with values 1, 2, 3, 4, 5 and 6. The object of the puzzle is to place below each of these cards a card from the set 7, 8, 9, 10, jack and queen so that the 12 numbers obtained by taking the sum and the absolute difference of each vertical pair of cards are all distinct. (The jack and the queen are assumed to have values of 11 and 12.)

For the case $n=7$ only one of the six basic patterns in only one of its four rotations generates a solution. This pattern and its mirror image generate two solutions to the number problem. Of the 12 basic patterns of nonattacking queens on the standard order-eight chessboard three have one rotation that avoids reflection attacks and another pattern has two. These five patterns and their mirror images generate 10 distinct solutions to the number problem in the case $n=8$. Interested readers can find all of them by exploring the rotations and reflections of the 12 basic order-eight patterns shown on page 192 of *The Unexpected Hanging.*

For square boards of orders higher than eight it is much easier to write a computer program for calculating the total number of solutions (including all rotations and reflections) to the variation on the nonattacking-queens problem than to write one that eliminates the rotations and reflections of the solutions to the basic problem. Some recent programs have, however, extended the number of basic patterns for orders nine through 16. The results obtained are respectively 46, 92, 341, 1,787, 9,233, 45,752, 285,053 and 1,846,955. The numbers for orders nine, 10 and 11 were correctly determined before computers came into wide use, but the higher-order results contradict all the claims and conjectures I have seen in older books.

Because the number of basic solutions increases rapidly as n increases it seems unlikely that none of them would furnish a solution when reflection strips are added. For this reason Klarner conjectures that for all n greater than 6 the number problem has at least one solution. In 1969 J. D. Sebastian reported on a computer program that found a solution to the number problem for every value of n from 9 through 27. The number of distinct solutions for values of n greater than 8 remains unknown.

At the start of a chess game the asymmetry introduced by the positions of the king and queen makes the pattern of white chessmen a mirror image of the pattern of black ones. Figure 100 shows how the Japanese graphic artist Mitsumasa Anno has extended this mirror-reflection symmetry to the players themselves as well as to the chessmen of a game. Because of

FIGURE 100 Mitsumasa Anno's mirror chess

Anno's passion for mathematical themes, he has been called the M. C. Escher of Japan. The illustration is reproduced from his book *Anno 1968–1977.*

Anno's picture calls to mind two whimsical chess tasks that Sam Loyd, America's greatest composer of puzzles and chess problems, published in 1866. In both tasks it is assumed that every move by Black will be a mirror-image duplicate of the preceding move by White. (Obviously we are concerned here only with possible, or legal, play, not with competent play.) The first task is to design a mirror-play game in which White checkmates on his fourth move. The second and much more difficult task is to design a mirror-play game in which White makes an eighth move that forces Black to checkmate him with a nonmirror move. (Students of chess problems call such a move a selfmate.)

How many different moves can be made on a standard chessboard? The answer is 1,840. Each move is represented by a line segment in the graph shown in Figure 101. The graph is taken from statistician I. J. Good's 1972 Christmas card. The 64 cells of the chessboard have been replaced by 64 dots, so that the graph appears to display a seven-by-seven board.

Good's graph calls to mind a nontrivial graph-coloring problem that has been applied to square chessboards. Given any one of the five different chess pieces, what is the minimum number of colors needed for coloring the cells of a board so that no matter where the piece is placed on the board it can only move to a cell of a different color? This minimum number is called the chromatic number of the order-n board for the piece.

The smallest chromatic number is 2, the number of colors needed for the knight on all boards of side three or more. The graph-coloring solution for any of these boards is simply to color its cells like those of a checkerboard. This finding gives a quick solution to the following brainteaser: If every cell of a chessboard is occupied by a knight, can all the knights simultaneously move to another cell? The answer is yes if and only if the board's side length is even, because only then are there equal numbers of cells of each color. If the side length is odd, there is an extra cell of one color. In that case, since every knight must move to a cell of the other color, there must be at least one knight with no place to go.

The chromatic number for the king on all square boards of side two or more is 4. If the four colors are labeled *A, B, C*

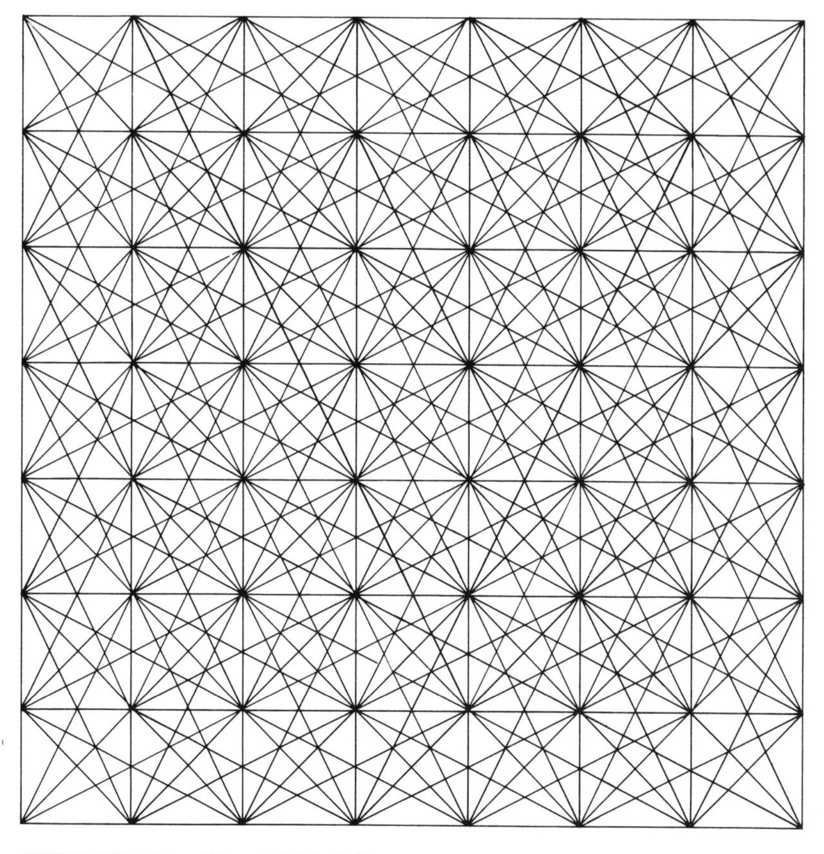

FIGURE 101 The 1,840 different moves in a chess game

and *D,* the top row of the board is colored *ABABAB* . . . , the second row *CDCDCD* . . . , the third row *ABABAB* . . . and so on. For a rook the chromatic number of an n-by-n board is n. To color the board begin at a corner and color parallel diagonal lines, using color *A* for the one cell of the first diagonal, *B* for the two cells of the second diagonal, *C* for the three cells of the third diagonal and so on until the nth color is used for the main diagonal. Starting again with *A* for the next diagonal, use the same sequence of colors for the remaining smaller diagonals.

For a bishop the chromatic number is n or $n-1$, depending on whether the longest diagonal on which the bishop can move has n or $n-1$ cells. If the bishop can move n cells, color each row a different color. If the bishop's longest move is $n-1$ cells, the color of the first row can be repeated on the nth row.

The case of the queen is the most interesting because it corresponds to the ancient combinatorial problem of forming "diagonal Latin squares," or n-by-n matrixes with cells colored so that no color is repeated within a single row, column or diagonal of any length. The minimum number of colors needed will obviously be equal to the chromatic number of an order-n chessboard for the queen. More than 60 years ago George Pólya proved that such n-by-n squares can be formed with n colors if and only if n is not divisible by 2 or 3. Pólya's method of coloring the five-by-five square is shown in Figure 102. Note that the hatched and dotted squares mark parallel chains of knight moves. (These chains are more evident in the pattern for the 11-by-11 square in the Dover reprint of Maurice Kraitchik's *Mathematical Recreations,* page 252.) Similar patterns generate solutions for any order-n board when n is not a multiple of 2 or 3.

Consider what happens on any such board if red queens are placed on all the red cells, blue queens on all the blue cells, green queens on all the green cells and so on. It will obviously be possible to place n differently colored sets of n queens on the n-by-n board in such a way that no queen of one color attacks another of the same color. The smallest board of this type (other than the trivial one-by-one board) is the order-five board. It has often been sold as a commercial puz-

FIGURE 102 A pandiagonal Latin square of order five

zle with 25 counters divided into five differently colored sets of five counters each. The task is to place the counters so that no counter shares a row, a column or a diagonal with another counter of the same color.

As far as I know the chromatic-number problem for the queen has not yet been solved for all order-n boards when n is a multiple of 2 or 3, although the answer is known for many low-order boards. For example, the six-by-six board requires seven colors and the eight-by-eight requires nine. It has been established that the queen's chromatic number, which cannot be less than n, cannot be greater than $n+3$.

Pólya's coloring method has an additional attribute that is often called the wraparound property: if either pair of opposite edges of the colored board are joined to make a cylinder, it will still be true that every diagonal contains n colors. In a different terminology each "broken diagonal" of the square has n colors. Squares with this property are called pandiagonal Latin squares.

A superqueen, or amazon, is a special chess piece that combines the moves of a queen and a knight. It is not possible to place n superqueens on an order-n board so that they do not attack one another when n is less than 10. The one solution in the case $n=10$ is shown in Figure 103. Solomon W. Golomb has proved that such patterns do exist when n is greater than 9 and is either a prime or one less than a prime. Ashok K. Chandra has shown that a wraparound pattern of n nonattacking superqueens exists for an order-n board if and

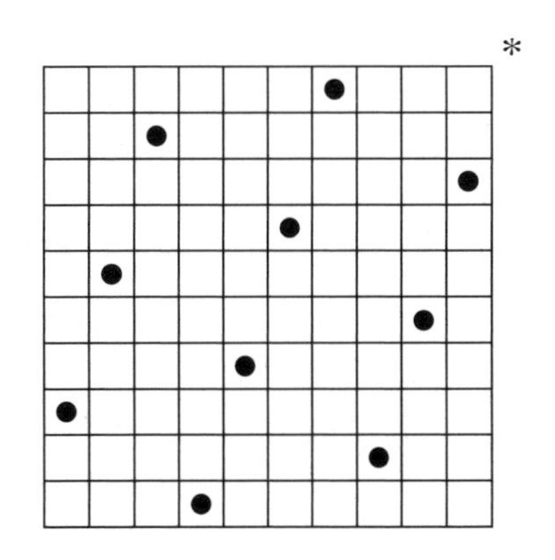

FIGURE 103 Smallest n-by-n board for n superqueens

only if n is greater than 11 and is not a multiple of 2 or 3. I know of no general solution of the superqueen problem.

Sin Hitotumatu of Kyoto wrote a computer program that found six solutions for the order-11 board, although all of them are identical if the board is wrapped around in both directions to make a torus. These solutions were found earlier by Golomb and are based on his construction method. To find one of them simply extend the order-10 solution shown in Figure 103 by adding a row at the top and a column at the right; the 11th superqueen is placed in the upper right-hand corner of the board, as is shown by the asterisk. Hitotumatu also found 22 basic solutions for the superqueen problem on the order-12 board. The one in Figure 104 is the prettiest, with its symmetrical arrangement of two equal squares and a smaller interior square. (The dotted lines show how the smaller square can be modified by rotation.)

Donald E. Knuth found a method for solving the order-14 and -15 cases that generalizes to certain higher orders. It is now known that if the Knuth and Chandra constructions are combined, solutions to the superqueen problem exist whenever n is greater than 9 and does not have the form $12k+8$ or $12k+9$ for some integer k. Chandra's method provides solutions for the cases $n=6k$, $n=6k+1$, $n=6k+4$ and $n=6k+5$; Knuth's method provides solutions for half of the remaining cases. Hence the smallest unresolved cases are $n=20$, $n=21$, $n=32$ and $n=33$.

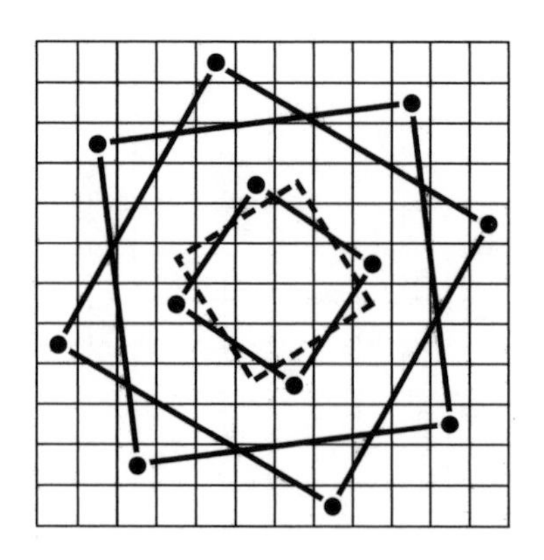

FIGURE 104 Twelve superqueens on a 12-by-12 board

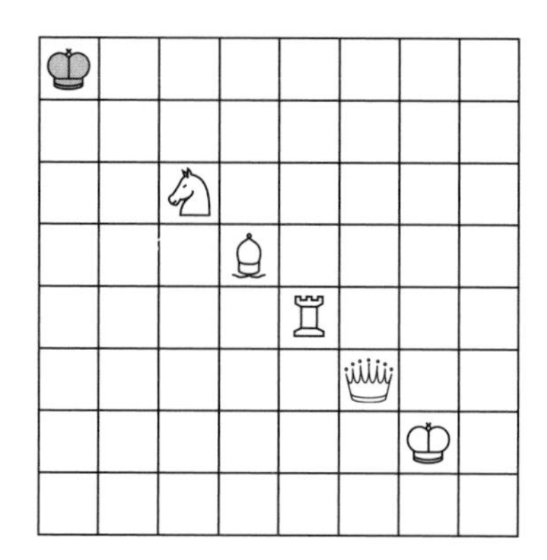

FIGURE 105 How can White mate in less than one move?

I shall conclude with a joke. How can White, in the position shown in Figure 105, checkmate Black by making a fraction of a move?

ANSWERS

Sam Loyd's mirror-move game, in which White mates on his fourth move, has two solutions:

	White	Black
1.	1. P-QB4	1. P-QB4
	2. Q-R4	2. Q-R4
	3. Q-B6	3. Q-B6
	4. QxB (mate)	
2.	1. P-Q4	1. P-Q4
	2. Q-Q3	2. Q-Q3
	3. Q-KR3 (or B5)	3. Q-KR3 (or B5)
	4. QxB (mate)	

Loyd's mirror-move game in which White self-mates on his eighth move is played as follows:

White	Black
1. P-K4	1. P-K4
2. K-K2	2. K-K2
3. K-K3	3. K-K3
4. Q-B3	4. Q-B3

5. N-K2	5. N-K2
6. P-QN3	6. P-QN3
7. B-R3	7. B-R3
8. N-Q4 (check)	

Black's only possible move is PxN, which checkmates the white king.

To achieve the joke checkmate in the final problem White simply raises his knight an inch or so above its square and shouts, "Discovered, mate!"

ADDENDUM

I said that the numbers of distinct solutions to the reflecting queens problem were not known beyond $n=8$. Paul Stevens, then at Madison, Wisconsin, was the reader whose computer program extended this the furthest. A chart of his results for n through 17 is given in Figure 106.

In 1980 I heard from three readers who established that solutions for nonattacking superqueens exist for any n greater than 9: Ashok Chandra at IBM; Charles Zimmerman of Madison, Wisconsin; and J. Reineke and P. Päppinghaus at the University of Hanover. I do not know if any such proofs have been published. Many readers extended the enumeration of

Side of Board	Solutions	Solutions with 90-Degree Rotational Symmetry
4	2	2
5	4	0
6	0	0
7	2	0
8	10	0
9	32	0
10	38	0
11	140	0
12	496	8
13	1,186	0
14	3,178	0
15	16,792	0
16	82,038	48
17	289,566	0

FIGURE 106 Paul Steven's ennumeration of distinct solutions to the reflecting queens problem

distinct solutions beyond $n = 12$, but again I know of no published results.

Numerous alternate solutions to the joke problem of mate in half a move were proposed by readers. Abraham Schwartz suggested lifting the bishop or queen slightly above the board, or moving the rook east and planting it on the line separating the rook from the knight column. Fred McCarthy had the ingenious thought of assuming that a white pawn on QB7 had moved to QB8 and had been removed, but the substitution of a queen for the pawn had not yet been made. He noted that the pawn could also have been on QR7 or QB7, and a black piece on QN1. The pawn has captured the piece and been removed, but not yet replaced by a mating queen or rook. If the captured black piece is on QB1, the white pawn can start on either QB7 or Q7 to be promoted to a queen after the capture. Similar tactics were sent by Freidrich-Wilhelm Scholz and Victor Feser. Feser said that strictly speaking, such mates should be called one-third-of-a-movemates because two steps have occurred—advancing the pawn, and its removal, with the replacement by a queen the third step. He cited earlier problems calling for mate in half a move in which the mate is achieved by half of a castling move.

BIBLIOGRAPHY

Sam Loyd and His Chess Problems. Alain C. White. Dover, 1962.

"Pairings of the First $2n$ Integers So That Sums and Differences Are All Distinct." J. L. Selfridge, in *Notes of the American Mathematical Society,* 10, 1963, page 195.

"Problem 1." Michael Slater, in *Bulletin of the American Mathematical Society,* 69, 1963, page 333.

"On Coloring the $n \times n$ Chessboard." M. R. Iyer and V. V. Menon, in *American Mathematical Monthly,* 73, 1966, pages 721–725.

"The Problem of Reflecting Queens." David Klarner, in *American Mathematical Monthly,* 74, 1967, pages 953–955.

"The Eight Queens and Other Chessboard Diversions." Martin Gardner, in *The Unexpected Hanging and Other Mathematical Diversions.* Simon and Schuster, 1969.

"Some Computer Solutions to the Reflecting Queens Problem." J. D. Sebastian, in *American Mathematical Monthly,* 76, 1969, pages 399–400.

"Independent Permutations, as Related to a Problem of Moser and a Theorem of Pólya." Ashok K. Chandra, in *Journal of Combinatorial Theory,* Series A, 16, 1974, pages 111–120.

"Generalized Latin Squares on the Torus." Henry D. Shapiro, in *Discrete Mathematics,* 74, 1978, pages 63–67.

"Toroidal *n*-Queens Problem." Solution to Problem E2698 by Richard Goldstein, in *American Mathematical Monthly,* 86, 1979, pages 309–310.

"Chess Tasks." Martin Gardner, in *Wheels, Life, and Other Mathematical Amusements,* Chapter 17. W. H. Freeman, 1983.

"Latin Squares and Superqueens." F. H. Hwang and Ko-Wei Lih, in *Journal of Combinatorial Theory,* Series A, 1983, pages 110–114.

16

Douglas Hofstadter's *Gödel, Escher, Bach*

This sentence no verb.

—DOUGLAS R. HOFSTADTER,
Gödel, Escher, Bach:
an Eternal Golden Braid

Every few decades an unknown author brings out a book of such depth, clarity, range, wit, beauty and originality that it is recognized at once as a major literary event. *Gödel, Escher, Bach: an Eternal Golden Braid,* a hefty (777 pages) volume published by Basic Books (1979), is such a work. The author (and the illustrator and typesetter) is Douglas R. Hofstadter, a young computer scientist at Indiana University who is the son of the well-known physicist Robert Hofstadter.

What can Kurt Gödel, M. C. Escher and Johann Sebastian Bach have in common? The answer is symbolized by the objects shown in the photograph that is Figure 107, and in the photograph on the book's jacket. In each photograph two wood blocks floating in space are illuminated so that their shadows on the three walls meeting at the corner of a room

FIGURE 107 *G*, *E* and *B* cast as shadows by a pair of "trip-lets"

form the initials of the three surnames Gödel, Escher and Bach. More precisely, the upper block casts *"GEB"* (Gödel, Escher, Bach) the heading of the book's first half, and the lower block casts *"EGB"* (Eternal Golden Braid) the heading of its second half. The letters *G*, *E* and *B* may be thought of as the labels for three strands that are braided by repeatedly switching a pair of letters. Six steps are required to complete a cycle from *GEB* (through *EGB*) back to *GEB*.

Dr. Hofstadter (his Ph.D. is in physics from the University of Oregon) calls such a block a "trip-let," a shortened form of "three letters." The idea came to him, he explains, "in a flash." Intending to write a pamphlet about Gödel's theorem, his thoughts gradually expanded to include Bach and Escher until finally he realized that the works of these men were "only shadows cast in different directions by some central solid essence." He "tried to reconstruct the central object, and came up with this book."

Hofstadter carved the trip-let blocks from redwood, using a band saw and an end mill. The basic idea behind this form is an elaboration of the classic puzzle that asks what solid shape will cast the shadows of a circle, a square and a triangle. Can a trip-let be constructed for any set of three different letters? If the letters may be distorted sufficiently, the answer is yes, and so to make the problem interesting some restrictions must be imposed. To begin with, the letters (preferably uppercase) must all be conventionally shaped, and they must fit snugly into the three rectangles that are the orthogonal projections of a rectangular block. In addition the solid must be connected, that is, it must not fall apart into separate pieces. It is not easy to determine except by trial and error whether such a trip-let can be made for any three given letters. As it turns out, some trip-lets are not possible. This problem suggests exotic variations, for example n-tup-lets that project n letters; four-dimensional tup-lets that project solid trip-lets that in turn project plane shadows of letters; solids that project numbers, pictures or words, and so on. (For this description of trip-lets I am indebted to Hofstadter's friend Scott Kim, who worked closely with him on many aspects of the book.)

What reality does Hofstadter see behind the work of his three giants? One aspect of that reality is the formal structure of mathematics: a structure that, as Gödel's famous undecidability proof shows, has infinitely many levels, none of which are capable of capturing all truth in one consistent system. Hofstadter puts it crisply: "Provability is a weaker notion than truth." In any formal system, rich enough to contain arithmetic, true statements can be made that cannot be proved within the system. To prove them one must jump to a richer system, in which again true statements can be made that cannot be proved, and so on. The process goes on forever.

Is the universe Gödelian in the sense that there is no end to the discovery of its laws? Perhaps. It may be that no matter how deeply science probes there will always be laws uncaptured by the theories, an endless sequence of wheels within wheels. Hofstadter argues eloquently for a kind of Platonism in which science, at any stage of its history, is like the shadow projections on the wall of Plato's cave. The ultimate reality is always out of reach. It is the Tao about which nothing can be said. "In a way," Hofstadter writes at the end of his preface, "this book is a statement of my religion."

For laymen I know of no better explanation than this book presents of what Gödel achieved and of the implications of his revolutionary discovery. That discovery concerns in particular recursion, self-reference and endless regress, and Hofstadter finds those three themes vividly mirrored in the art of Escher, the most mathematical of graphic artists, and in the music of Bach, the most mathematical of the great composers. The book's own structure is as saturated with complex counterpoint as a Bach composition or James Joyce's *Ulysses*. The first half of the book serves as a prelude to the second, just as a Bach prelude introduces a fugue. Moreover, each chapter is preceded by a kind of prelude, which early in the book takes the form of a "Dialogue" between Achilles and the Tortoise. Other characters enter later: the Sloth, the Anteater, the Crab and finally Alan Turing, Charles Babbage and the author himself. Each Dialogue is patterned on a composition by Bach, and in several instances the mapping is strict. For example, if the composition has n voices, so does the corresponding Dialogue. If the composition has a theme that is turned upside down or played backward, so does the Dialogue. Each Dialogue states in a comic way, with incredible wordplay (puns, acrostics, acronyms, anagrams and more), the themes that will be more soberly explored in the chapter that follows.

There are two main reasons for Achilles and the Tortoise having been chosen to lead off the Dialogues. First, they play the major roles in Zeno's paradox (the topic of the book's first Dialogue), in which Achilles must catch the Tortoise by escaping from an infinite regress. Second, they are the speakers in an equally ingenious but less familiar paradox devised by Lewis Carroll. In Carroll's paradox, which Hofstadter reprints as his second Dialogue, Achilles wishes to prove Z, a theorem of Euclid's, from premises A and B. The Tortoise, however, will not accept the theorem until Achilles postulates a rule of inference C, which explicitly states that Z follows from A and B. Achilles adds the rule to his proof, thinking the discussion is over. The Tortoise then, however, jumps to a higher level, demanding another rule of inference D, which states that Z follows from A, B and C, and so it goes. The resulting endless regress seems to invalidate all reasoning in much the same way that Zeno's paradox seems to invalidate all motion. "Plenty of blank leaves, I see!" exclaims Carroll's Tortoise, glancing at Achilles' notebook. "We shall need them ALL!" The warrior shudders.

One of Hofstadter's Dialogues, "Contracrostipunctus," is an acrostic (complete with punctuation marks) asserting that if the words in it are taken backward, they provide a second-order acrostic spelling out "J. S. Bach." Another Dialogue, "Crab Canon," which is illustrated with an Escher periodic tessellation of crabs, is based on Bach's "Crab Canon" in his *Musical Offering.* As the Tortoise discusses Bach, his sentences are interspersed with those of Achilles, who is discussing Escher. The Tortoise and Achilles use the same sentences in reverse order. The Crab enters briefly at the crossing point to knot the halves of their discourse together, halves that interweave in time in the same way that the positive and negative crabs of Escher's tessellation interweave in space.

The initials *A, T* and *C* (for Achilles, Tortoise and Crab) correspond to the initials of adenine, thymine and cytosine, three of the four nucleotides of DNA, the molecule with the extraordinary ability to replicate itself. Just as Achilles pairs with the Tortoise, so adenine pairs with thymine along the DNA double helix. Cytosine pairs with guanine. The fact that the initial *G* can be taken to stand for "gene" prompted Hofstadter to do a "little surgery on the Crab's speech" so that it would reflect this coincidence. The striking parallel between the tenets of mathematical logic and the "central dogma" of molecular biology is dramatized in a chart Hofstadter calls the "Central Dogmap."

The letter *G* also stands for Gödel's sentence: the sentence at the heart of his proof that asserts its own unprovability. To Hofstadter the sentence provides an example of what he calls a Strange Loop, exemplifying the self-reference that is one of the book's central themes. (A framework in which a Strange Loop can be realized is called a Tangled Hierarchy, and the letters of "sloth" turn out to stand for "Strange Loops, or Tangled Hierarchies.") Dozens of examples of Strange Loops are discussed, from Bach's endlessly rising canon (which modulates to higher and higher keys until it loops back to the original key) to the looping flow of water in Escher's *Waterfall* and the looping staircase of his *Ascending and Descending.* One of the most amusing models of *G* is a record player X that self-destructs when a record titled "I Cannot Be Played on Record Player X" is played on it.

A particularly striking example of a two-step Strange Loop is Escher's drawing of two hands, each one sketching the other. We who see the picture can escape the paradox by "jumping out of the system" to view it from a metalevel, just as we can escape the traditional paradoxes of logic by jumping into a metalanguage. We too, however, have Strange Loops, because the human mind has the ability to reflect on itself, that is, the firing of neurons creates thoughts about neurons. From a broader perspective the human brain is at a level of the universe where matter has acquired the awesome ability to contemplate itself.

By the end of *Gödel, Escher, Bach* Hofstadter has introduced his readers to modern mathematical logic, non-Euclidean geometries, computability theory, isomorphisms, Henkin sentences (which assert their own provability), Peano postulates (the pun on "piano" is not overlooked), Feynman diagrams for particles that travel backward in time, Fermat's last theorem (with a pun on "fermata"), transfinite numbers, Goldbach's conjecture (which is cleverly linked with Bach's *Goldberg Variations*), Turing machines, computer chess, computer music, computer languages (Terry Winograd, an expert on the computer simulation of natural language, appears in one Dialogue under the anagrammed name of Dr. Tony Earrwig), molecular biology, the "mind" of an anthill called Aunt Hillary, artificial intelligence, consciousness, free will, holism v. reductionism, and a kind of sentence philosophers call a counterfactual.

Counterfactuals are statements based on hypotheses that are contrary to fact, for example, "If Lewis Carroll were alive today, he would greatly enjoy Hofstadter's book." These statements pose difficult problems in the semantics of science, and there is now a great deal of literature about them. For Hofstadter they are instances of what he calls slipping, progressing from an event to something that is almost a copy of it. The Dialogue that precedes a chapter on counterfactuals and artificial intelligence concerns a Subjunc-TV set that enables an observer to get an "instant replay" of any event in a football game and see how the action would have looked if certain parameters were altered, that is, if the ball were spherical, if it were raining, if the game were on the moon, if it were played in four-dimensional space and so on.

The book's discussion of artificial intelligence is also enormously stimulating. Does the human brain obey formal rules of logic? Hofstadter sees the brain as a Tangled Hierarchy: a multilevel system with an intricately interwoven and deep self-referential structure. It follows logical rules only on its molecular substrate, the "formal, hidden, hardware level" where it operates with eerie silence and efficiency. No computer, he believes, will ever do all a human brain can do until it somehow reproduces that hardware, but he has little patience with the celebrated argument of the Anglican philosopher J. R. Lucas that Gödel's work proves a human brain can think in ways that are in principle impossible for a computer.

Only a glimpse can be given here of the recreational aspects of this monstrously complicated book. In "The Magnificrab, Indeed" (a pun on Bach's *Magnificat* in D), the Dialogue that introduces a discussion of deep theorems of Alonzo Church, Turing, Alfred Tarski and others, appears a whimsical Indian mathematician named Mr. Najunamar. Najunamar has proved three theorems: he can color a map of India with no fewer than 1,729 colors; he knows that every even prime is the sum of two odd numbers, and he has established that there is no solution to $a^n + b^n = c^n$ when n is zero. All three are indeed true.

Some readers will recognize 1,729 as the number of the taxi in which G. H. Hardy rode to visit the Indian mathematician Srinivasa Ramanujan ("Najunamar" spelled backward) in a British hospital. Hardy remarked to Ramanujan that 1,729 was a rather dull number. Ramanujan rejoined instantly that on the contrary it was the smallest positive integer that is the sum of two different pairs of cubes. Hardy then asked his friend if he knew the smallest such number for fourth powers. Ramanujan did not know the number, although he guessed that it would turn out to be fairly big. Hofstadter supplies the answer: 635,318,657, or $134^4 + 133^4$ or $158^4 + 59^4$. He also wonders if his readers can find the smallest number that can be expressed as the sum of two squares in two different ways, but he hides the answer. Can you determine it before I supply it in the answer section?

To explain the meaning of the term "formal system" Hofstadter opens his book with a simple example that uses only the symbols *M, I* and *U.* These symbols can be arranged in strings called theorems according to the following rules:

1. If the last letter of a theorem is *I, U* can be added to the theorem.

2. To any theorem *Mx, x* can be added. (For example, *MUM* can be transformed into *MUMUM,* and *MU* can be transformed into *MUU.*)

3. If *III* is in a theorem, it can be replaced by *U,* but the converse operation is not acceptable. (For example, *MIII* can be transformed into *MU,* and *UMIIIMU* can be transformed into *UMUMU.*)

4. If *UU* is in a theorem, it can be dropped. (For example, *UUU* can be transformed into *U,* and *MUUUIII* can be transformed into *MUIII.*)

There is only one "axiom" in the system: In forming theorems one must begin with *MI.* Every string that can be made by applying the rules, in any order, is a theorem of the system. Thus *MUIIU* is a theorem because it can be generated from *MI* in six steps. If you play with the *M, I* and *U* system, constructing theorems at random, you will soon discover that all theorems begin with *M* and that *M* can occur nowhere else.

Now for a puzzle: Is *MU* a theorem? I shall say no more about *MU* here except that it plays many other roles in the book, in particular serving as the first two letters of "Mumon," the name of a Zen monk who appears in a delightful chapter on Zen koans.

Even as simple a system as that of *M, I* and *U* enables Hofstadter to introduce a profound question. If from all the possible strings in the system we subtract all the strings that are theorems, we are left with all the strings that are not theorems. Hence the "figure" (the set of theorems) and the "ground" between the theorems (the set of nontheorems) seem to carry equivalent information. Do they really? Is the system like an Escher tessellation in which the spaces between animals of one kind are animals of another kind, so that reproducing the shapes of either set automatically defines the other? (Or so that a black zebra with white stripes is the same as a white zebra with black stripes?) In this connection Hofstadter reproduces a remarkable tessellation by Kim in which the word "FIGURE" is periodically repeated in black so that the white ground between the black letters forms the same shapes [see Figure 108]. The same concept is playfully illustrated in the Dialogue "Sonata for Unaccompanied Achilles" (modeled on Bach's sonatas for unaccompanied violin), in which we hear

FIGURE 108 Scott Kim's FIGURE FIGURE Figure. From *Inversions* (W. H. Freeman & Co., 1989).

only Achilles' end of a telephone conversation with the Tortoise about "figure" and "ground." From Achilles' half of the conversation we can reconstruct the Tortoise's.

Other figure-ground examples are provided by the counting numbers. For example, given all the primes, we can determine all the nonprimes simply by removing the primes from the set of positive integers. Is the same true of all formal systems? Can we always take all the theorems from the set of all possible statements in the system and find that what is left—the set of nontheorems—is another complementary formal system? An unexpected discovery of modern set theory is that this is not always the case. To put it more technically, there are recursively enumerable sets that are not recursive. Thus does Hofstadter lead his readers from trivial beginnings into some of the deepest areas of modern mathematics.

The book closes with the wild Dialogue "Six-Part Ricercar," which is simultaneously patterned after Bach's six-part ricercar and the story of how Bach came to write his *Musical Offering*. (A ricercar is a complicated kind of fugue.) In this Dialogue the computer pioneers Turing and Babbage improvise at the keyboard of a flexible computer called a "smart-stupid," which can be as smart or as stupid as the programmer wants. (The computer's name is a play on "pianoforte," which means "soft-loud.") Turing produces on his computer screen a simulation of Babbage. Babbage, however, is seen looking at the screen of his own smart-stupid, on which he has conjured up a simulation of Turing. Each man insists he is real and the other is no more than a program. An effort is made to resolve the debate by playing the Turing Game, which was proposed by Turing as a way to distinguish a human being from a computer program by asking shrewd questions. The conversation in this scene parodies the conversation Turing gives in his classic paper on the topic.

At this point Hofstadter himself walks into the scene and convinces Turing, Babbage and all the others that they are creatures of his own imagination. He, however, is as unreal as any of the other characters of the Dialogue, because he too is imagined by the author. The situation resembles a painting by René Magritte titled *The Two Mysteries,* in which a small picture of a tobacco pipe is displayed with a caption that says (to translate from the French) "This is not a pipe." (see Figure 109.) Floating above the fake pipe is a presumably genuine larger pipe, but of course it too is painted on the canvas.

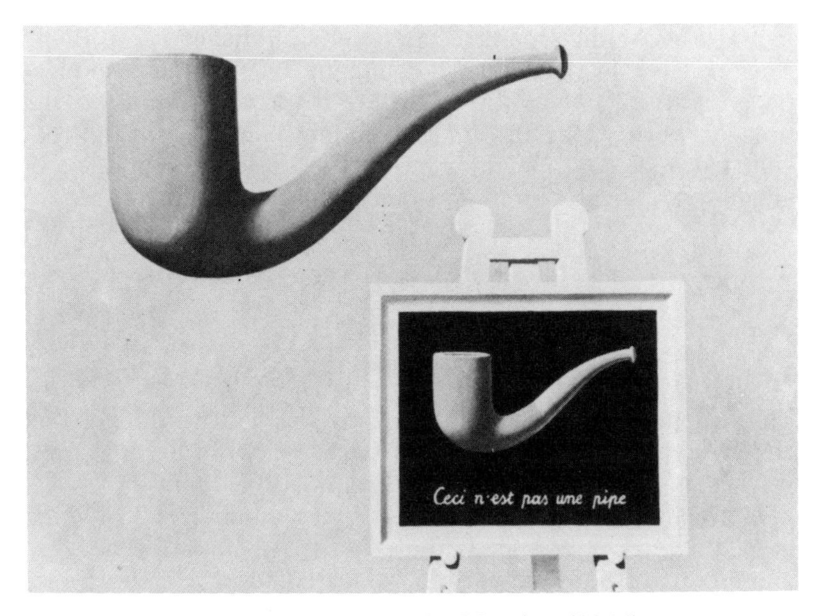

FIGURE 109 *The Two Mysteries,* by Magritte (1966)

And how real was Magritte? How real are Hofstadter, you and I? Are we in turn no more than shadows on the rapidly flipping leaves of the book we call the universe? We are back to a Gödelian Platonism in which reality is infinitely layered. Who can say what reality really is? The book's final word, "RICERCAR," is a multilevel pun anticipated by a series of acronyms such as Turing's remark "Rigid Internal Codes Exclusively Rule Computers and Robots" as well as Bach's inscription "Regis Iustu Cantio et Reliqua Canonica Arte Resoluta" on a sheet of music he sent to Frederick the Great. Bach's six-part ricercar is from his *Musical Offering,* the story of which opens the book. In this way RICERCAR serves, in much the same way as "riverrun," the first word of *Finnegans Wake,* to twist the work into one gigantic self-referential loop.

One of the book's most amusing instances of a Strange Loop is in the Dialogue that introduces Chapter 16. The Crab speaks of browsing through a "crackpot" book on "metal-logic" titled *Copper, Silver, Gold: an Indestructible Metallic Alloy.* Hofstadter's annotated bibliography reveals that this book is by one Egbert B. Gebstadter (note the *EGB* of Egbert, the *GEB* of Gebstadter and the *EBG* initials) and was published in 1979 by Acidic Books (Hofstadter's publisher being Basic Books). Here is Hofstadter's comment: "A formidable hodgepodge, turgid and confused—yet remarkably similar to the present work. Contains some excellent examples of indirect self-reference. Of particular interest is a reference in its well-annotated bibliography to an isomorphic, but imaginary, book."

ANSWERS

The first problem was to find the smallest positive integer that can be expressed as the sum of two squares in two different ways. The number is 50, which equals 5^2+5^2 or 1^2+7^2. If zero squares are allowed, however, the number is 25, which equals 5^2+0^2 or 3^2+4^2. If the two squares must be nonzero and different, the solution is 65, which equals 8^2+1^2 or 7^2+4^2.

The second problem was to determine whether or not *MU* is a theorem in the *M, I* and *U* formal system. A simple proof of why *MU* is not a theorem can be found on pages 260 and 261 of *Gödel, Escher, Bach: an Eternal Golden Braid.*

ADDENDUM

Hofstadter's *GEB* took off like a rocket, staying long on the best seller lists, and winning the 1980 Pulitzer Prize for general nonfiction. Vintage Books paid $200,000 for paperback rights—the largest sum it had ever paid for nonfiction rights, and the largest sum Basic Books ever received for such a work.

Reviews in 1979 were lavish in their praise. Especially noteworthy were reviews by Brian Hayes (*New York Times Book Review,* April 29), Walter Kerrick (*Village Voice,* November 19) and Edward Rothstein (*New York Review of Books,* December 6). Other reviews ran in *Commonweal, Technology Review, Psychology Today, American Scientist, Yale Review, American Scholar,* and *New Republic.*

In an amusing review in the *Journal of Recreational Mathematics* (14, 1981–82, pp. 52–54), Leon Bankoff observed that

GEB has exactly 777 pages, and by using the cipher $A=1$, $B=2$, $C=3$, and so on, one discovers that $G=E+B$.

Hofstadter became my successor in writing the Mathematical Games column in *Scientific American,* after he changed the department's name to Metamagical Themas, an anagram of its former title. His columns were reprinted in *Metamagical Themas: Questing For the Essence of Mind and Pattern* (Basic Books, 1985), a work of 852 pages. A few years earlier, Hofstadter and Daniel C. Dennett had edited a marvelous anthology, *The Mind's I: Fantasies and Reflections on Self and Soul* (Basic Books, 1981). At present Hofstadter is professor of cognitive science and computer science and technology at the Indiana University in Bloomington.

I must confess that I could never have written my review of *GEB* had I not had on hand a 33-page analysis of the book written by Scott Kim titled *Strange Loop Gazette.* Kim has since obtained his doctorate under Donald Knuth, in the computer science department of Stanford University. Kim's beautiful book *Inversions,* containing scores of names and phrases drawn in such a way that they magically remain the same when inverted or mirror reflected (or turn into another word or phrase) has been reissued by W. H. Freeman. (A book of similar inversions by Hofstadter, titled *Ambigrammi,* was published in Italy in 1987 but has yet to have a U.S. edition.) Kim is now working on a book about how to use one's fingers to model such things as the skeleton of a cube or tetrahedron, or to entwine the fingers to produce such topological structures as a trefoil knot.

I showed how *MUIIU* could be generated from MI in six steps. Several readers lowered this to five, and one reader, Raymond Aaron, did it in four: *MI* to *MII* (rule 2), to *MIIII* (rule 2), to *MIIIIIIII* (rule 2), and finally to *MUIIU* (rule 3).

I did not give a proof that *MU* is not a theorem. Here is how Hofstadter handled it. Every theorem begins with *M,* which occurs nowhere else. The number of I's in a theorem is not a multiple of 3 because this is true for the axiom *MI,* and every permissible operation preserves this property. Therefore *MU,* whose number (zero) of I's is a multiple of 3, cannot be obtained by the permissible operations.

Several readers wrote programs for determining Ramanujan numbers that solve the Diophantine equation $A^n+B^n=C^n+D^n$. When n is 3, William J. Butler, Jr.'s program found 4,724 solutions for values less than 10^{10}, of which the largest is

$$1,956^3 + 1,360^3 = 2,088^3 + 964^3.$$

Of the 4,724 solutions, 26 are triples, the smallest being

$$414^3 + 255^3 = 423^3 + 228^3 = 436^3 + 167^3.$$

The number of primitive solutions (no common factor of the four numbers) is infinite, but the number of triples, Butler conjectures, could be finite because their density declines rapidly as the numbers grow in size.

Hofstadter disclosed in a letter that the phrase "formidable hodgepodge" in his joke review of *GEB* (quoted in my final paragraph) was taken from a reviewer's comment when Indiana University considered publishing the book. (The book was also rejected, incidentally, by an editor then at W. H. Freeman.) Because W. V. Quine, the Harvard philosopher, was one of the two reviewers, the chances are fifty percent that the phrase was Quine's. "Turgid and confused" (the phrase also appears on p. 3 of *GEB*) is from a comment on Bach's style by one of his pupils. In *GEB*'s second printing Hofstadter added the following to his hoax review: "Professor Gebstadter's Shandean digressions include some excellent examples of indirect self-reference." The first four words, with the change of name, are from Brian Hayes's *New York Times* review of *GEB*.

BIBLIOGRAPHY

"Exploring the Labyrinth of the Human Mind." James Gleick, in the *New York Times Magazine,* August 21, 1983. A coverstory on Douglas Hofstadter.

"The Copycat Project." George Johnson, in *Machinery of the Mind: Inside the New Science of Artificial Intelligence,* Chapter 15. Times Books, 1986. The chapter is about Hofstadter's approach to artificial intelligence.

17

Imaginary Numbers

"The imaginary numbers are a wonderful flight of God's Spirit; they are almost an amphibian between being and not being."

—LEIBNIZ

In a column on negative numbers that is reprinted in my *Penrose Tiles to Trapdoor Ciphers,* I described how long it took and how painful it was for mathematicians to extend the definition of "number" to include negative numbers. The same process was repeated with even more anguish when mathematicians discovered the enormous usefulness of what unfortunately were named imaginary numbers. It is a strange and beautiful story.

Although there were a few early pronouncements that negative quantities cannot have square roots (because the square of any real number must be positive), the story of imaginary numbers really begins in 16th-century Europe. At that time mathematicians, in particular Rafaello Bombelli of Italy, found that in solving algebraic problems it was often useful to assume that negative numbers did have square roots. In other words, just as the equation $x+1=0$ could be solved only by setting x equal to -1, so $x^2+1=0$ could be solved only by setting x equal to $\sqrt{-1}$.

The seemingly preposterous assumption that there is a square root of -1 was justified on pragmatic grounds: it simplified certain calculations and so could be used as long as "real" values were obtained at the end. The parallel with the rules for using negative numbers is striking. If you are trying to determine how many cows there are in a field (that is, if you are working in the domain of positive integers), you may find negative numbers useful in the calculation, but of course the final answer must be in terms of positive numbers because there is no such thing as a negative cow.

Throughout the 17th and 18th centuries mathematicians everywhere kept discovering new uses for the square roots of negative numbers. It was Leonhard Euler who in the 18th century introduced the symbol i (the first letter of the Latin word *imaginarius*) for $\sqrt{-1}$. A much-quoted statement attributed to Euler asserts that such roots are not nothing or more than nothing or less than nothing but strictly imaginary or impossible. Mathematicians eventually worked out the algebraic rules for manipulating the "pure imaginaries" (the products of i and real numbers) and what later came to be called complex numbers (the sums of pure imaginaries and real numbers).

A complex number has the form $a+bi$, where a and b can be any real numbers. (In this instance the plus sign is not meant to indicate addition in the familiar sense; it serves mainly to separate the real part a of the complex number from its imaginary part bi.) In other words, if a equals 0 and b does not equal 0, the complex number is a pure imaginary bi. If b equals 0, then bi drops out, leaving the real number a. Therefore, the complex numbers include as subsets all the reals and all pure imaginaries, just as the real numbers include all the integers, fractions and irrationals. In modern terminology the complex numbers form the mathematical structure called a field, whose elements obey all the familiar laws of arithmetic. The complex-number field is closed with respect to addition, subtraction, multiplication and division, that is, applying those operations to any two complex numbers will always generate another number in the field. There is a sense in which the discovery of the complex field completes traditional algebra because it makes possible the solution of any ordinary algebraic equation whatsoever. The field also turned out to be closed with respect to the operations employed in the calculus, and that discovery gave rise to a vast edifice of mathematics concerning the functions of a complex variable.

Many of the advances of modern physics could not have been made without the extension of algebra to the complex field. The first major scientific use of complex numbers was made by Charles Proteus Steinmetz, who found them essential for doing efficient calculations on alternating currents. Today no electrical engineer could get along without them, and neither could any physicist working on the area of air or fluid dynamics. The numbers also play a basic role in relativity theory (where space-time is made symmetrical by the stratagem of regarding the three spatial dimensions as real and the time dimension as imaginary), in quantum mechanics and in many other branches of modern physics.

Because there are still misgivings about calling i a number, it is not uncommon even today for a physicist, a philosopher or even a mathematician to maintain that i is not really a number but is only a symbol for an operation I shall explain below. No one has disposed of this verbal quibble more effectively than Alfred North Whitehead. In the chapter on imaginary numbers in his *Introduction to Mathematics* he wrote:

> At this point it may be useful to observe that a certain type of intellect is always worrying itself and others by discussion as to the applicability of technical terms. Are the incommensurable numbers properly called numbers? Are the positive and negative numbers really numbers? Are the imaginary numbers imaginary, and are they numbers?—are types of such futile questions. Now, it cannot be too clearly understood that, in science, technical terms are names arbitrarily assigned, like Christian names to children. There can be no question of the names being right or wrong. They may be judicious or injudicious; for they can sometimes be so arranged as to be easy to remember, or so as to suggest relevant and important ideas. But the essential principle involved was quite clearly enunciated in Wonderland to Alice by Humpty Dumpty, when he told her, apropos of his use of words, "I pay them extra and make them mean what I like." So we will not bother as to whether imaginary numbers are imaginary, or as to whether they are numbers, but will take the phrase as the arbitrary name of a certain mathematical idea, which we will now endeavour to make plain.

Complex numbers behave so much like ordinary numbers when they are added, subtracted, multiplied and divided (according to the rules of the complex field) that most mathematicians no longer hesitate to call them numbers and regard them as having just as much "reality" as negative numbers. Even the counting numbers are no more than symbols manipulated according to the rules of a deductive system. We think of them as being more "real" than other numbers only because their applications are so close to our practical experience of counting fingers, cows, people and so on. What we forget is that only the fingers, cows and people are real, not the symbols to which we turn to count them. In the realm of pure mathematics i is just as real as 2. If we like, we can think of 2 as nothing more than an operator: a symbol that tells us to double 1.

Most people are so accustomed to working with real numbers, however, that they feel great relief when they discover there is a simple geometrical interpretation of complex numbers. This interpretation, which makes it easy to "see" what the numbers are all about, identifies every complex number with a point on the Cartesian plane. The first person to make this ingenious connection was Caspar Wessel, a self-taught Norwegian surveyor who lectured on it in 1797. A few years later the idea was rediscovered by Jean-Robert Argand, a Swiss bookkeeper (who published a small book about it in 1806), and independently by the great German mathematician Carl Friedrich Gauss.

As is shown in Figure 110, the basic idea is to view the horizontal axis of the Cartesian plane as the real-number line and the vertical axis as the line of points that correspond to the pure imaginary numbers. In other words, one-to-one correspondences are established between the real numbers and the points on the x axis and between the pure imaginary numbers and the points on the y axis. As I have pointed out, both of these sets can be considered subsets of complex numbers, and now the remaining complex numbers can be put in one-to-one correspondence with the remaining points on the plane. To obtain the coordinates of the point associated with a complex number, one simply measures the real part on the real axis and the imaginary part on the imaginary axis. The points corresponding to four complex numbers are shown in the illustration.

With this interpretation of the complex numbers, it is possible to forget entirely the disturbing notion that i is the

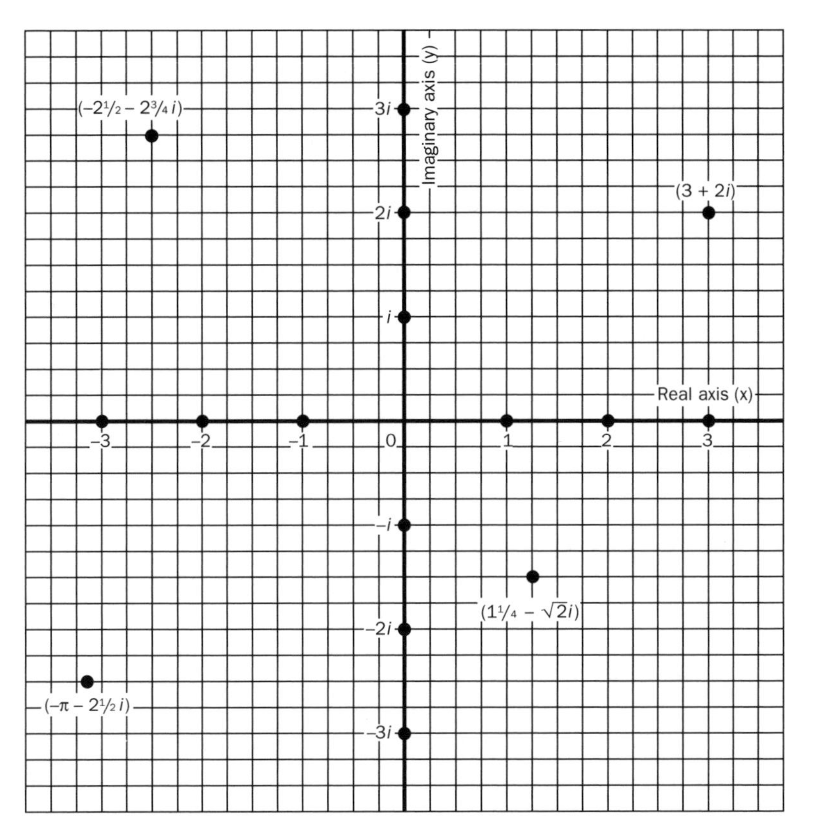

FIGURE 110 Correspondence of complex numbers with points on the complex plane

square root of -1 (which of course it is not in the usual sense of taking a square root). Now a complex number can be viewed simply as an ordered pair of real numbers: the first number measured on the real axis and the second on the imaginary axis. In other words, by properly defining the arithmetic operations for combining these pairs, it is possible to construct an algebra of ordered pairs of real numbers that is equivalent to the algebra of complex numbers. That opaque phrase, "the square root of a negative number," is nowhere encountered in this new algebra, although the same idea is of course present in a different language and in different notation. If this algebra of ordered pairs had been developed before complex numbers, perhaps today no one would remember imaginary numbers and wonder whether or not they exist.

After the discovery of this geometrical interpretation of complex numbers, mathematicians immediately asked whether the basic concept could be generalized to three dimensions, that is, to points in space, or, to put it another way, to ordered triples? The answer, alas, is no—not without a radical modification of the laws of arithmetic. As Eric Temple Bell once put it, the complex field is "the end of a road." It was the Irish mathematician William Rowan Hamilton who made the first breakthrough into "hypercomplex numbers" when he invented quaternions: four-part numbers that combine a real number with three imaginaries. The key to manipulating them is the fact that they do not obey the commutative law for multiplication: the rule stating that for any two numbers a and b, ab equals ba.

The idea of dropping this law came to Hamilton in 1843 as he was strolling with his wife at dusk along the Royal Canal in Dublin. He was so elated that he stopped to scratch the basic formula on a stone of Brougham (pronounced "broom") Bridge. The original graffiti weathered away in Hamilton's time, and now the bridge is known only as the one that crosses Broombridge Street. There is a tablet in the stone commemorating the great event, however, and in 1943, a century after Hamilton's revelation, Ireland honored it with a postage stamp. Quaternions do not form a field (their structure is called a division ring), but the algebra of quaternions is equivalent to an algebra of ordered quadruplets and is often applied today as a part of three-dimensional vector theory. The discovery of the algebra of quaternions marked the beginning of modern abstract algebra, in which all kinds of "numbers", much stranger than the complex numbers, can be defined.

Because of the correspondence between complex numbers and points on the Cartesian plane, when the plane is used in this way it is called the complex plane. (It is also called the z plane, for the unspecified complex number z equals $a+bi$, and sometimes an Argand diagram, because for many decades no one knew about Wessel's earlier discovery.) I shall not go into the details of how complex numbers can be added, subtracted, multiplied and divided by geometrical diagrams on the complex plane. Readers who do not already know the rules governing these operations can find them in any elementary algebra textbook that covers complex numbers. A brief explanation of multiplication by i, however, is necessary to introduce an elegant theorem about the roots of numbers.

To multiply a number on the complex plane by i, one takes the radius-vector line to the point corresponding to the number (the line from the origin of the plane to the point) and rotates it 90 degrees counterclockwise; the new end point of the vector corresponds to the product of the number and i. It is in such a sense that i can be viewed as an operator. To understand this idea, consider what happens when i is raised to various powers: i raised to the first power of course equals i, and it is easy to see that i^3 equals -1, i^3 equals $-i$ and i^4 equals 1. This four-step cycle repeats endlessly: i^5 equals i, i^6 equals -1, i^7 equals $-i$, i^8 equals 1 and so on. All even powers of i equal 1 or -1, and all odd powers equal i or $-i$.

Figure 111 shows how these observations apply to the multiplication of a number (in this case pi, or π) by i. The point corresponding to π on the positive side of the x axis is located and moved 90 degrees counterclockwise around the circle of radius π centered at the origin of the plane. An arrow shows how the end point of this operation is the pure imaginary, πi, which lies on the upper part of the y axis. Multiplying π by i^2, then, is equivalent to multiplying it twice by i: the point corresponding to π is moved 180 degrees around the circle and ends up at the point $-\pi$ on the x axis, or real-number line. Similarly, multiplying π by i^3 effects a turn of 270 degrees, ending at the point $-\pi i$ on the lower part of the y axis; multiplying π by i^4 is the same as multiplying π by 1, and so we are back to π. We can continue in the same way with all the higher powers of i. Each next-highest power takes us another quarter turn counterclockwise around the circle.

The inverse operation of multiplication by i is division by i: moving clockwise 90 degrees around the center origin of the plane. In other words, for any complex number, draw a radius-vector line from the origin to the point that represents the number. Then, to multiply the number by i, rotate the vector 90 degrees counterclockwise (see Figure 112), and to divide it by i, rotate the vector 90 degrees the other way. (As a joke, a friend of mine once suggested that i times infinity equals 8 because multiplying by i turns the infinity sign upright.)

With this interpretation of multiplication, it turns out that if complex roots are counted, every nonzero number (real or complex) has exactly n nth roots. In other words, every num-

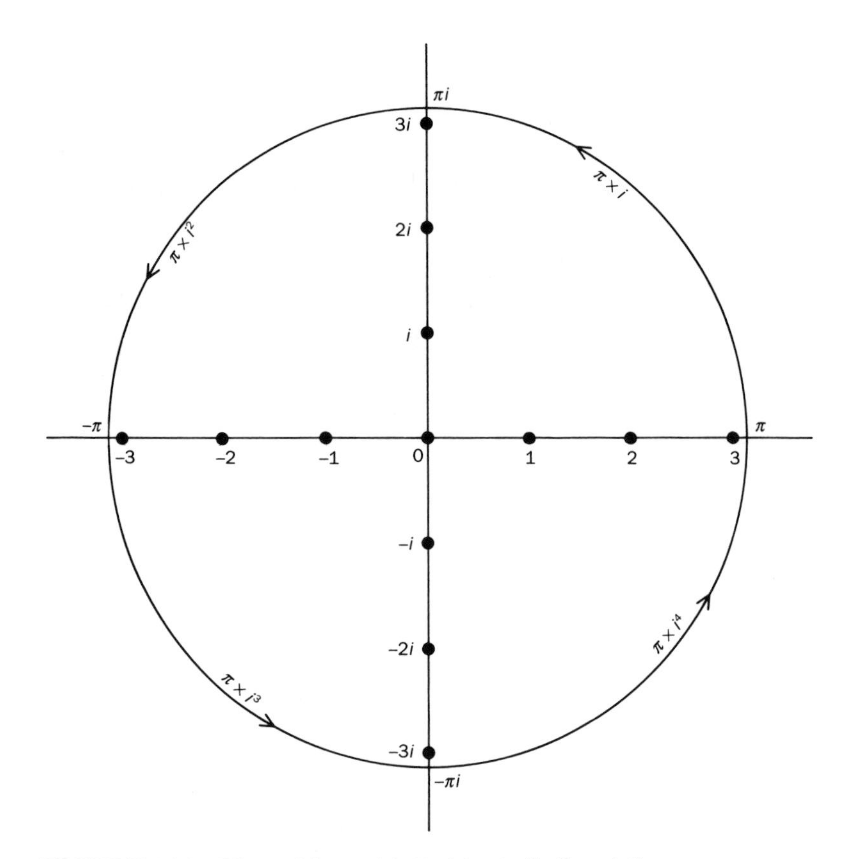

FIGURE 111 How pi is multiplied by i, i^2, i^3 and i^4

ber has two square roots, three cube roots, four fourth roots, five fifth roots and so on. It follows that every cubic equation has three solutions, every fourth-degree equation has four solutions and so on, and when we diagram the roots of individual numbers on the complex plane, an unexpected and delightful property is revealed. The n points corresponding to the nth roots all lie, separated by equal distances, on a circle whose origin is at the center of the plane. In other words, the points mark the corners of a regular n-sided polygon. For example, Figure 113 shows the locations of the six sixth roots of 729. If, as in this instance, the number is real and has an even number of roots, two corners of the polygon will lie on the real axis. If the number is real and has an odd number of roots, only one corner of the polygon will lie on the real axis.

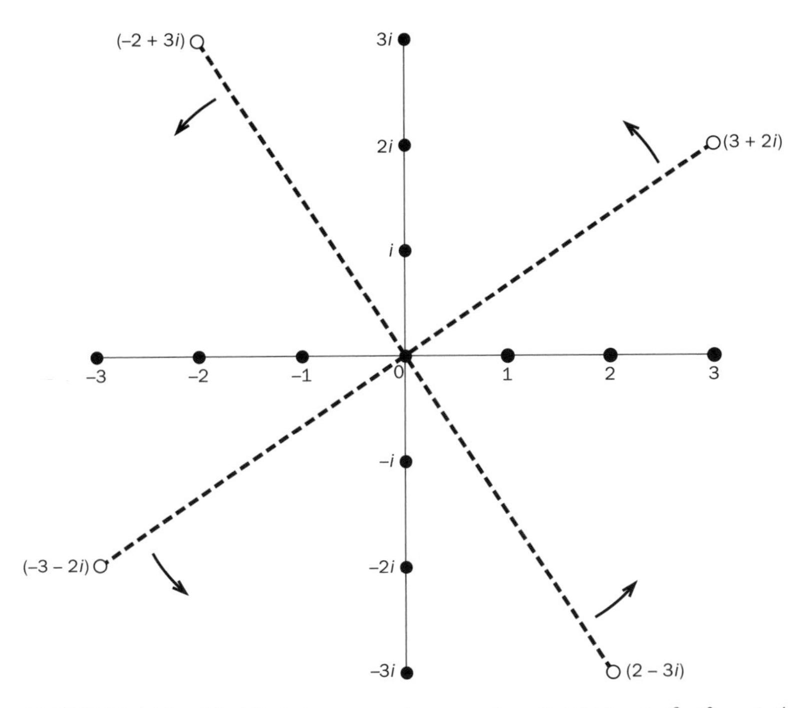

FIGURE 112 Multiplying complex number $3+2i$ by i, i^2, i^3 and i^4

If n is a rational number, not an integer, there is a finite number of roots that correspond to points on the circle. If n is irrational, the roots still lie on the circle but are infinite in number as you go around the circle forever. Given all real numbers, their roots catch all points on the circle, but for a particular irrational, an infinity of "holes"—points not covered—remain.

Besides being essential in modern physics, the complex-number field provides pure mathematics with a multitude of brain-boggling theorems. It is worth keeping in mind that complex numbers, although they include the reals as a subset, differ from real numbers in startling ways. One cannot, for example, speak of a complex number as being either positive or negative: those properties apply only to the reals and the pure imaginaries. It is equally meaningless to say that one complex number is larger or smaller than another.

It had been known before Euler that the product of any two pure imaginaries is a real number, but it was Euler who

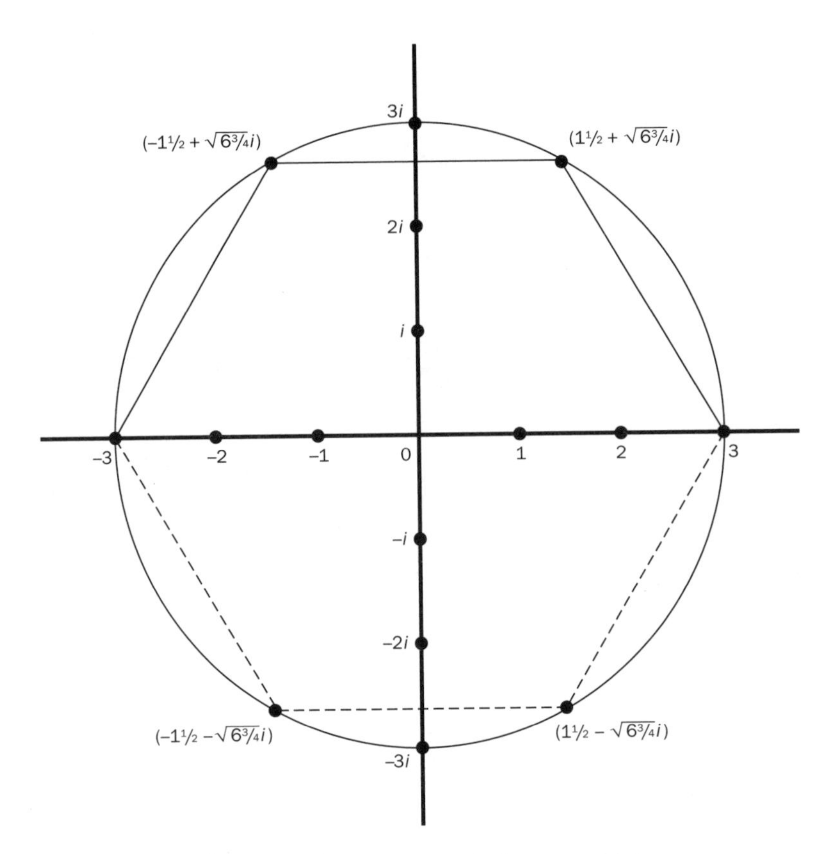

FIGURE 113 The six sixth roots of 729

first showed that i^i is also real. It is equal to $e^{-\pi/2}$, an irrational number with the decimal expansion of .2078795763. . . . Actually this number is only one of an infinity of values, all real, of i^i. They are given by the formula $e^{-\pi/2 \pm 2k\pi}$, where k is any integer, so that when k equals 0, the principal value given above is obtained. The ith root of i is also a real number, the principal value of which is $e^{\pi/2}$, or 4.8104773809. . . .

There are many other formulas in which i is related to the two best-known transcendental irrationals, e (the natural basis of logarithms) and π. The most famous formula, developed by Euler but based on an earlier discovery, is $e^{i\pi}+1=0$, which

Edward Kasner and James Newman call in their book *Mathematics and the Imagination* "elegant, concise and full of meaning." They also quote a remark by Benjamin Peirce, the Harvard mathematician who was the father of Charles Sanders Peirce, about the formula: "Gentlemen," he said, after writing the formula on a blackboard, "that is surely true, [but] it is absolutely paradoxical; we cannot understand it, and we don't know what it means, but we have proved it, and therefore we know it must be truth."

Well, the formula is not quite without meaning. Rewritten as $e_i^{\pi} = -1$, it can be diagrammed on the z plane as the limit of the infinite sequence: $1 + \pi i + (\pi i)^2/2! + (\pi i)^3/3! + (\pi i)^4/4! \ldots$ (The exclamation point is the factorial sign: $n!$ equals $1 \times 2 \times 3 \times \ldots \times n$.) The terms of this sequence are diagrammed as an infinite set of points on a counterclockwise spiral of straight lines that strangles the -1 point on the real axis.

George Gamow, seeking to dispel the mystery of complex numbers, once devised this puzzle. An old parchment, describing the location of buried pirate treasure on a desert island, gave the following instructions. On the island there are only two trees, *A* and *B*, and the remains of a gallows. Start at the gallows and count the steps required to walk in a straight line to tree *A*. At the tree, turn 90 degrees to the left and then walk forward the same number of steps. At the point where you stop, drive a spike into the ground. Now return to the gallows and walk in a straight line, counting your steps, to tree *B*. When you reach the tree, turn 90 degrees to the right and take the same number of steps forward, placing another spike at the point where you stop. Dig at the point exactly halfway between the spikes, and you will find the treasure.

A young adventurer who found the parchment with these instructions chartered a ship and sailed to the island. He had no difficulty finding the two trees but, to his dismay, the gallows was gone and time had abolished all traces of where it had stood. Not knowing the location of the gallows, he could see no way of finding the treasure and so returned empty-handed. Gamow points out that if the young man had been familiar with the technique of manipulating numbers on the complex plane, he could have found the treasure with ease. Readers who know the basic rules for diagramming complex numbers should be able to solve this problem.

ANSWER

George Gamow's problem of finding buried treasure can be solved as follows even if one does not know the location of the gallows. Draw a straight line through trees A and B, as is shown in Figure 114, and call the line the real axis of the complex plane. Then through the point midway between the trees, draw the perpendicular imaginary axis. Consider that tree A is at the point representing the real number 1, and tree B is at the point representing the real number -1. Choose any point as the location of the gallows.

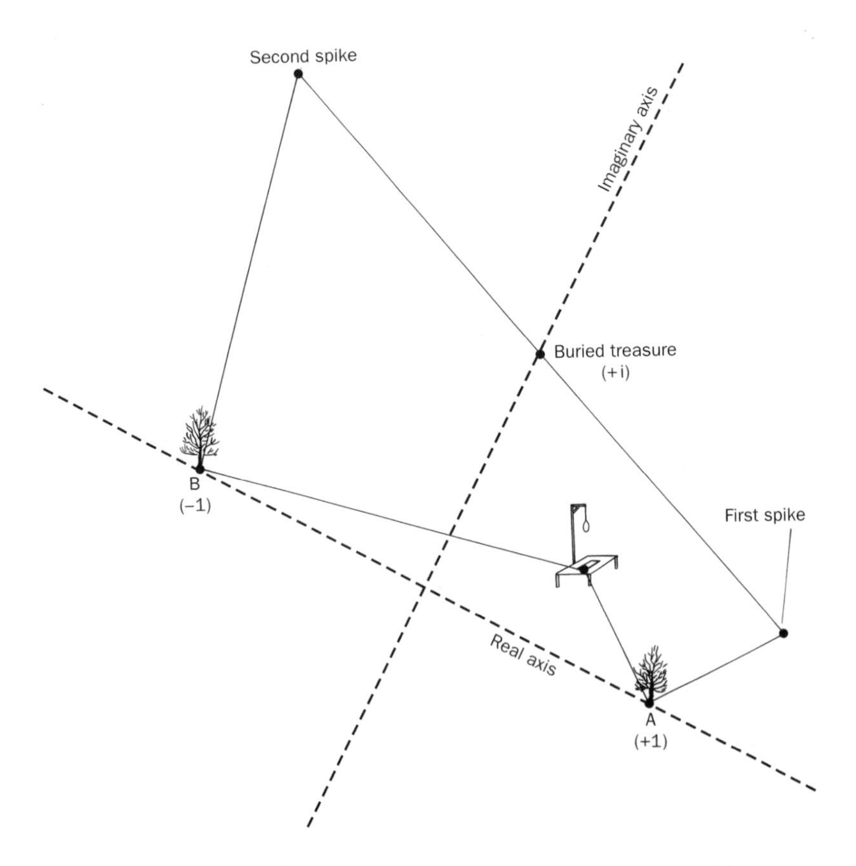

FIGURE 114 Solution to the buried-treasure problem

Now, if the instructions on the parchment are followed, it becomes clear that, no matter where the gallows stood, the treasure will be at $+i$ on the imaginary axis! This observation can be proved easily by applying the rules for the manipulation of numbers on the complex plane. The problem appears in the second chapter of Gamow's popular book *One Two Three—Infinity*. Readers who want to check the proof will find it explained clearly there.

ADDENDUM

Because every complex number corresponds to a vector on the plane, Gamow's problem can be solved by vector geometry without using complex numbers. Many readers have sent me such solutions. Keith Raybourn cracked the problem by first assuming it had a solution. Since we are not told where the gallows used to be, it follows that its position could have been anywhere. One can then locate the treasure by selecting an aribitrary spot for the gallows and following instructions.

F. V. Pohle, of Adelphi University, sent me a copy of a 1977 West German stamp honoring Gauss that depicts four numbers on the complex plane (see Figure 115). Enlarged, it would make a handsome mathematical poster.

As far as I know, the last physicist to oppose the use of complex numbers was the eccentric Swiss physicist Baron Stueckelberg (see the bibliography for an article about him), who died in 1984. He is almost unknown today, yet he was

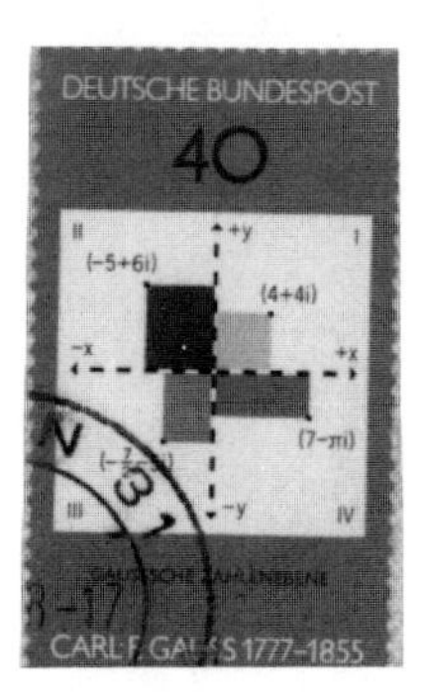

FIGURE 115 This German stamp, honoring Gauss, shows the location of four points on the complex plane

the first to discover the strong force and made several notable contributions that led to Nobel prizes by others. Stueckelberg spent years trying to persuade his colleagues to stop using imaginary numbers in the equations of relativity and quantum theory. It is true that physicists could get by, if they had to, by abandoning imaginary numbers. They could even dispense with irrational numbers because every measurement of a magnitude is rational. Irrational constants, such as pi and *e* and the square root of 2, can always be expressed to a finite number of decimal places that are adequate for any calculation. But dropping the use of irrationals and imaginary numbers would be as foolish and inefficient as trying to dispense with negative numbers.

BIBLIOGRAPHY

"Convergence on the Argand Diagram." L. W. H. Hull, in *The Mathematical Gazette,* 43, 1959, pages 205–207.

Complex Numbers in Geometry. I. M. Yaglom. Academic Press, 1968.

"The Historical Development of Complex Numbers." D. R. Green, in *The Mathematical Gazette,* 60, 1976, pages 99–107.

"Hamilton's Discovery of Quaternions." B. L. van der Waerden, in *Mathematics Magazine,* 49, 1976, pages 227–234.

"Line Reflections in the Complex Plane—a Billiard Player's Delight." Gary L. Musser, in *Mathematics Teacher,* 71, 1978, pages 60–64.

"The Physicist That Physics Forgot: Baron Stueckelberg's Brilliantly Obscure Career." Robert P. Crease and Charles C. Mann, in *The Sciences,* July–August, 1985, pages 18–22.

18

Pi and Poetry: Some Accidental Patterns

"A mathematician, like . . . a poet, is a maker of patterns."

—G. H. HARDY,
A Mathematician's Apology

Are the beautiful, orderly patterns of pure mathematics discovered or created by the human mind? The answer to the question depends on one's philosophy of mathematics. In either case, patterns also play an important part in all the fine arts, and nature displays a fantastic variety of patterns—atomic structures, snowflakes, spiral galaxies and so on—that demonstrate a collusion between mathematical and natural laws. In all these domains, however, patterns can arise completely by chance, as a cloud can assume the shape of a camel. In this chapter I shall take a not so serious look first at some accidental patterns involving the numbers pi and *e* and then at some whimsical instances of accidental poetry.

Pi is the best-known of the transcendental numbers: those irrational numbers that are not roots of ordinary algebraic equations. Given a sufficient amount of computing time, one can precisely calculate the decimal expansion of pi to any finite length, so in this sense the expansion is not random. Viewed as a sequence of digits, however, it is as ugly and disordered as any randomly generated list of numbers. No one has ever found a pattern in the expansion of pi that cannot be explained by pure chance. Nevertheless, this disheveled sequence, which was calculated in 1988 in Japan to 201,326,000 decimal digits, continues to haunt and fascinate numerologists. It is not surprising that with diligent searching one can discover all kinds of accidental patterns there. Dr. Matrix has pointed out a few of them, as I have reported in previous books. There are other pi curiosities. For example, on the title page of Dr. Matrix' 10-volume commentary on the King James Bible he quotes Job 14:16: "Thou numberest my steps" (see Chapter 18 of *The Magic Numbers of Dr. Matrix*). Here Dr. Matrix apparently overlooked a remarkable coincidence: "Job" has three letters, and adding 14:16 gives 3.1416, or pi rounded to the fourth decimal place.

Starting with the 710,100th decimal digit of the expansion of pi there is a surprising run of seven consecutive 3's: . . . 353733333338638. . . . (Note that the sequence is preceded by a 7 and followed by an 8.) Such stupendous expansions became possible only recently, as a result of the design of faster algorithms for computer multiplication and the development by Eugene Salamin of a new formula for calculating pi. By adding some clever twists to Karl Friedrich Gauss's method of calculating elliptic integrals, Salamin obtained a formula for pi that converges with unusual rapidity. Interested readers will find the formula explained in Salamin's 1976 paper cited in the bibliography.

The October 1965 issue of *Eureka,* a journal put out annually by mathematics students at the University of Cambridge, points out a strange pattern in the first seven decimal digits of pi. The pattern combines the three mystic numbers of medieval numerology (1 for the Godhead, 3 for the Trinity and 7 for the day God rested) with the first three perfect numbers (the smallest integers equal to the sum of all their divisors, including 1) as follows: the first decimal digit of pi is the smallest perfect number 1, the first three decimal digits (141) add up to the second perfect number (6) and the first seven decimal digits (1415926) add up to the third perfect number (28). Moreover, 1 is the "sum" of the first counting number, 6 is the sum of the first three counting numbers (1+2+3), and 28 is the sum of the first seven counting numbers. These are the only three numbers that are simultaneously the sum of the first *n* counting numbers and the sum of the first *n* decimal digits of pi.

I have written before about that extraordinary fraction 355/113, which gives the value of pi to six decimal places. G. Stanley Smith discovered that 533/312 is a good approximation of the square root of pi, 1.7724538509 . . . , giving it correctly to four decimal places. Even more remarkable, Smith's fraction is almost 355/113 with the numerator and denominator written backward. Note that the denominator of Smith's fraction begins with 31. The cube root of 31 gives pi to three decimal places. The square root of 9.87 (a number consisting of three consecutive digits in reverse order) is still better. It gives pi rounded to four decimal places.

The second most famous transcendental number is the base of natural logarithms, *e*, which has been the subject of almost as much numerology as pi. The expansion of *e* is 2.718281828. . . . The repetition of 1828 in this sequence means absolutely nothing. And the fact that the 16th and 17th decimal digits of pi and *e* are the same (23) is equally meaningless. Douglas R. Hofstadter has discovered an even more amazing coincidence: if the reciprocals of the first eight counting numbers are added (taking each one in decimal form and rounding up to the third decimal place when the fourth-place digit is greater than 5), the result is 2.718, or *e* to three decimals (see Figure 116).

1/1 =	1.000
1/2 =	.500
1/3 =	.333
1/4 =	.250
1/5 =	.200
1/6 =	.167
1/7 =	.143
1/8 =	.125
	2.718

FIGURE 116 Do these fractions add up to *e*?

Considering the fundamental nature of pi and e, it is not surprising that there are many simple, meaningful formulas that relate them, such as the Euler formula. As someone once observed, "You can have your pi and e at it too." It is also not surprising, however, that if one searches long enough, it is possible to find striking but meaningless relations between the two constants. One of the best was discovered recently by R. G. Duggleby, a biochemist at the University of Ottawa. He found that the sum of pi to the fourth power (97.40909 . . .) and pi to the fifth power (306.01968 . . .) is e to the sixth power (403.42879 . . .) correct to four decimal places!

Proving which is larger, pi to the power of e or e to the power of pi, is an old but still intriguing problem. (Obviously the proof may not involve the calculation of the two values, which are quite close: π^e equals 22.4591577 . . . , and e^π equals 23.1406926. . . .) Dozens of proofs have already been published. One of the shortest is based on the fact from elementary calculus that $x^{1/x}$ has a maximum value when x equals e. Hence $e^{1/e}$ is greater than $\pi^{1/\pi}$. Multiplying each exponent by π^e and canceling yields the inequality $e^\pi > \pi^e$. Incidentally, e^π has been proved to be transcendental, but it is not yet known whether π^e is rational or irrational. It is not even known whether the product (πe) and the sum $(\pi + e)$ of the numbers are rational or irrational.

If x is larger than y, then all positive rational solutions to the equation $x^y = y^x$ are given by the formulas $x = (1 + 1/n)^{n+1}$ and $y = (1 + 1/n)^n$, where n is any positive integer. Setting n as being equal to 1 generates the only solution in positive integers: $4^2 = 2^4$. There is one other solution in integers. Can you find it?

Figure 117 shows the graph of the equation $x^y = y^x$, when x and y are positive real numbers. The straight line gives all the solutions when x and y are equal, and the curve (which looks like a hyperbola but is not) gives all the solutions when x and y are not equal. On one side of the axis of symmetry x is greater than y, and on the other side it is less than y. The asymptotes of the curve are shown as broken lines. This function has been generalized to negative numbers, complex numbers and even transfinite numbers.

Do the first n digits of pi (which form the sequence 3, 31, 314, 3141, . . .) ever make a perfect square? This curious question is discussed by Wolfgang Haken in his 1977 paper. (The four-color theorem was proved in 1976 by Haken and

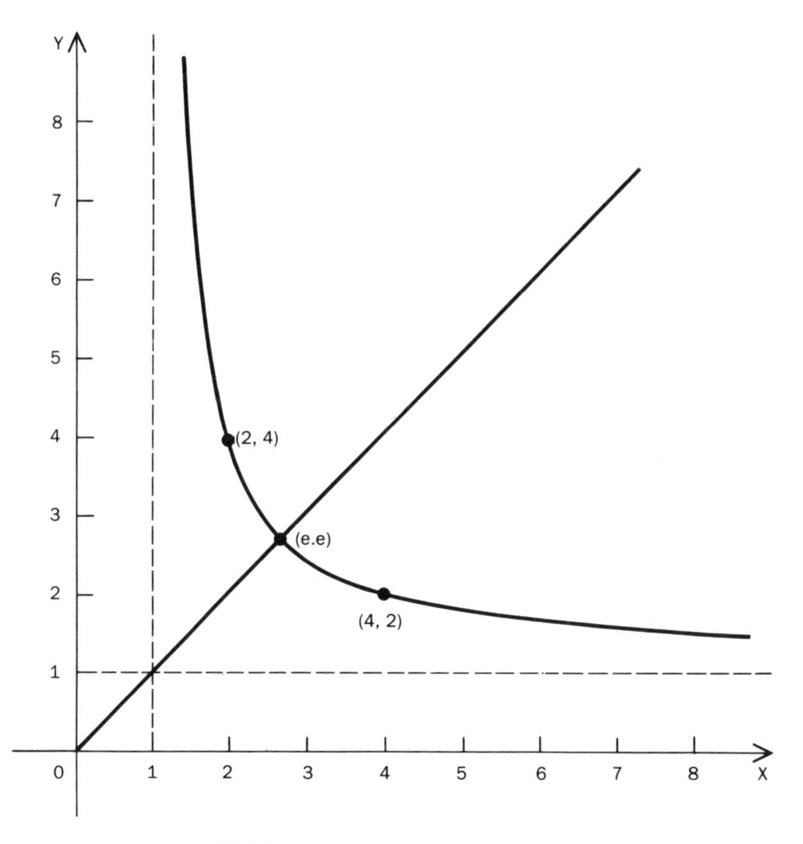

FIGURE 117 Graph of $x^y = y^x$

Kenneth Appel.) Consider the conjecture that the answer is no, the first *n* digits of pi never yield a perfect square. Haken believes this conjecture is true but not provable in standard set theory because "the decimal expansion of the transcendental number pi has 'practically nothing' to do with perfect squares." Haken estimates the probability that the conjecture is false to be .000000001.

Another bizarre question about pi was raised in a letter from George Shombert, Jr., of Beaver, Pa. If the decimal expansion of pi is truly patternless, then somewhere in the infinite sequence of digits there must be the first *n* digits of *e*. Similarly, it must be possible to find the first *n* digits of pi in the expansion of *e*. This observation set me wondering. Is it possible to prove that there is no point inside pi where *e* begins and continues to infinity, or vice versa? Is it possible that each of these numbers contains all of the other, as in 3.141 . . . 2718 . . . 3141 . . . 2718 . . . ? The answer to this last question is definitely no. Do you see why?

One of the most incredible accidental patterns involving pi was discovered a few years ago by T. E Lobeck of Minneapolis. He started with a conventional 5-by-5 magic square shown at the left in Figure 118, and then substituted the *n*th digit of pi for each number *n* in the square. The result is the matrix shown at the right in the illustration. The sum of the numbers in each row is shown to the left of the row, and the sum of the numbers in each column is shown at the bottom of the column. Amazingly, every column sum duplicates a row sum.

It may be hard to believe this pattern is sheer coincidence. Even mathematicians can forget that if enough people doodle long enough with random sequences of digits, it is highly probable they will find highly improbable patterns. It is because most people fail to grasp this basic notion that they are unduly impressed when, out of the billions upon billions of possible ways coincidences can arise in daily life, one does occur. As Edgar Allan Poe wrote at the beginning of his story "The Mystery of Marie Roget": "There are few persons, even among the calmest thinkers, who have not occasionally been startled into a vague yet thrilling half-credence in the super-

17	24	1	8	15
23	5	7	14	16
4	6	13	20	22
10	12	19	21	3
11	18	25	2	9

24	2	4	3	6	9
23	6	5	2	7	3
25	1	9	9	4	2
29	3	8	8	6	4
17	5	3	3	1	5
	17	29	25	24	23

FIGURE 118 Traditional magic square (left) is transformed by pi (right)

natural, by *coincidences* of so seemingly marvelous a character that, as *mere* coincidences, the intellect has been unable to receive them."

Just as it is not surprising to find patterns in disorderly sequences such as the expansions of pi and *e,* so it is not surprising, considering the vast quantity of word sequences that are published as prose, to find examples of accidental verse. I am referring not to free verse but to verse with an orderly pattern of meter or rhyme, preferably both. It is trivially easy to take a passage of purple prose and break it into lines that give it the semblance of poetry. That has often been done with passages from the King James Bible and even from the novels of Charles Dickens. When William Butler Yeats compiled *The Oxford Book of Modern Verse,* he included a "poem" on the Mona Lisa he had found in an essay by Walter Pater. Similarly, in 1945 John S. Barnes published a book called *A Stone, A Leaf, A Door,* which consists of prose passages from the novels of Thomas Wolfe presented in the form of verse.

About a dozen years ago there was a minor flurry of interest in "found poetry" comparable to the enthusiasm for such faddish "found art" as pieces of driftwood. At that time the British weekly *The New Statesman* ran a competition for the best found poems. Simon and Schuster published Ronald Gross's *Pop Poems,* a collection of free verse Gross had found in cookbooks, advertisements, insurance policies, tax forms, newspaper obituaries and other equally unlikely places. In Canada, several anthologies of found free verse were published by the poet John Robert Colombo. In John Updike's *Telephone Poles and Other Poems,* you will find a "poem" he came on in the prose of James Boswell's *Life of Samuel Johnson.*

In addition to excluding found free verse from consideration, I shall exclude the poems authors sometimes hide intentionally in their prose. For example, Washington Irving concealed 22 lines of iambic pentameter in the first paragraph of the sixth book of his *History of New York,* written under the pseudonym Diedrich Knickerbocker. (The poem begins: "But now the war-drum rumbles from afar. . . .") Lewis Carroll liked to write letters to little girls that appeared to be prose but were actually rhymed and in meter, and his apparent prose introduction to a long poem, "Hiawatha's Photographing," has the same tom-tom beat as the poem. James Branch Cabell occasionally concealed poems in his novels, and F. Scott Fitzgerald's *This Side of Paradise* has many italicized passages of word painting that turn out to be rhymed poems when they are read properly. Now and then a journalist will write a newspaper story in verse and present it as prose. For instance, John Canaday once reviewed an art show in *The New York Times* (December 27, 1964) with a full page of prose that turned out to be made up entirely of rhymed couplets.

By accidental verse, then, I mean lines that have the unintentional structure of rhymed poetry. The most famous English example is in William Whewell's *An Elementary Treatise on Mechanics* (1819). It came to light during a dinner in Whewell's honor at Cambridge, where Whewell was Master of Trinity College. Adam Sedgwick, a geologist, rose and asked if anyone knew who had written the following stanza:

And hence no force, however great,
Can stretch a cord, however fine,
Into a horizontal line
That shall be absolutely straight.

Although the same stanza form had been used by Tennyson in "In Memoriam," no one could identify those particular lines. Sedgwick then revealed that he had quoted from page 44 of the first volume of Whewell's treatise, taking the liberty to polish the last line, which originally read, "Which is accurately straight." Whewell was so unamused that he altered the passage to eliminate the verse in the next edition of the book. He later published a volume of poetry called *Sunday Thoughts and Other Verses,* but his accidental verse is the only one he wrote that is still remembered.

A splendid specimen of chance doggerel is found in Lincoln's second inaugural address:

Fondly do we hope,
Fervently do we pray,
That this mighty scourge of war
May speedily pass away.
Yet, if it God wills
That it continue until . . .

William Harmon, in his first book of poetry, *Treasury Holiday,* points out that the Declaration of Independence begins with the two rhyming words "When in," the Constitution begins "We the," and the Gettysburg Address begins "Fourscore." Harmon began his second book of poetry with "We see," and his introduction to the forthcoming *Oxford Book of American Light Verse* begins "Those o's." Dickens' *Bleak House,* Harmon reminds me, starts with the two-syllable sentence: "London."

There are countless other examples of accidental verse in all major languages. Here are two I stumbled on myself in *The New York Times.* James Thurber ended his article "The Quality of Mirth" (February 21, 1960) as follows:

If they are right and we are wrong,
I shall return to the dignity
of the printed page where it may be
That I belong.

And the first paragraph of James Reston's column "Mr. Ford's Last Chance" (January 16, 1976) ends with

This is the sound of prominent men,
Prodded by their wives,
Cleaning out the attic
And fleeing for their lives.

A kind of inverse of the unintended poem is the planned poem that, as the result of an accident of language, acquires a flaw that spoils a line. Consider this uncouth quatrain from "Daisy," an otherwise lovely lyric by Francis Thompson:

Her beauty smoothed earth's furrowed
face!
She gave me tokens three:—
A look, a word of her winsome mouth,
And a wild raspberry.

And finally, the first version of Robert Frost's "Maple" contained the line, "Her mother's bedroom was her father's still." After an alert proofreader caught the unintended meaning, Frost changed the last word to "yet."

Answers

Readers were asked to find a solution in integers of $x^y = y^x$, with x greater than y, other than the familiar 4, 2. The second solution is -2, -4.

Another exercise was to prove that it is impossible for the decimal expansions of pi and e to contain all of each other. If it were possible then, after a finite number (x) of decimal digits in pi, e would begin, and after a finite number (y) of decimal digits in e, pi would begin. In that case both pi and e would be repeating decimal fractions, each with a period equal to $x+y$. All repeating decimal fractions are rational, however, and since pi and e are known to be irrational, the assumption must be false.

BIBLIOGRAPHY

"Metric Prose." Charles C. Bombaugh, in *Oddities and Curiosities of Words and Literature.* Dover, 1961.
"Prose Poems." William S. Walsh, in *Handy-Book of Literary Curiosities.* Lippincott, 1904. Gale reprint, 1966.
Pop Poems. Ronald Gross. Simon and Schuster, 1967.
"Found Poetry." Ronald Gross, in *New York Times Book Review,* June 11, 1967, page 2.
"Graphical Solution of the Equation $a^b=b^a$." Thomas W. Shilgalis, in *The Mathematics Teacher,* 66, 1973, page 235.
"Computation of π Using Arithmetic Geometric Mean." Eugene Salamin, in *Mathematics of Computation,* 30, 1976, pages 565–570.
"An Attempt to Understand the Four Color Problem." Wolfgang Haken, in *Journal of Graph Theory,* 1, 1977, pages 193–206.
"Prose Poems." Tony Augarde, in *The Oxford Guide to Word Games.* Oxford University Press, 1984, pages 147–148.
"Slicing Pi Into Millions." Martin Gardner, in *Gardner's Whys and Wherefores,* Chapter 9. University of Chicago Press, 1989.

19

More on Poetry

In giving instances in the previous chapter of verse intentionally disguised as prose, of prose containing accidental verse, and of verse damaged by accidental meanings, I barely raked the surface. Moreover, I left out many amusing examples because *Scientific American* is a family magazine.

Regarding intentionally concealed poetry, William Harmon, a poet and professor of English literature at the University of North Carolina, called my attention to the title page and introduction, both concealing rhyme and meter, of James Russell Lowell's *A Fable for Critics* (1849). His kinswoman Amy Lowell later used the same whimsical device for the title page and introduction of *A Critical Fable* (1922).

In one of his essays in *The Sewanee Review* (reprinted in paperback, *Uneeda Review, 23½ Anniversary Issue* (Nick Lyons

Books, 1984, p. 20), Harmon added a footnote that seems to be in prose. Read aloud, it turns into the following limerick:

There once was a Thompson named Stith
Who reduced every item of myth,
Song, fable, and mummery
To a seven-word summary:
No extraneous details, just pith.

Several readers reminded me of Mark Twain's story "Punch, Brother, Punch," based on an 1874 poem by Isaac H. Bromley, in turn said to be based on a sign that Bromley saw on a horse-drawn streetcar:

The conductor when he receives a fare
Will punch in the presence of the passinjare,
A blue trip-slip for an 8-cent fare,
A buff trip-slip for a 6-cent fare,
A pink trip-slip for a 3-cent fare,
All in the presence of the passinjare.
Punch, brother, punch, punch with care,
Punch in the presence of the passinjare.

Other readers recalled the lyrics of a song, sung to the tune of *Humoresque,* and apparently based word for word on a sign that once hung in Pullman car washrooms:

Passengers will please refrain
From flushing toilets while the train
Is standing in the station, ...

The last line has such variant terminations as "Ah, how true!" "The hell with you!" and "I love you."

On the copyright page of *The Works of Max Beerbohm* are the lines: "London: John Lane, THE BODLEY HEAD. *New York: Charles Scribner's Sons."* In an essay on Beerbohm, John Updike tells us that Max had scribbled below these lines: "This plain announcement, nicely read, iambically runs." Updike comments:

> The effortless a-b-a-b rhyming, the balance of "plain" and "nicely," the need for nicety in pronouncing "Iambically" to scan—this is quintessential light verse, a twitting of the starkest prose into perfect form, a marriage of earth with light, and quite magical. Indeed, were I a high priest of literature, I would have this quatrain made into an amulet and wear it about my neck, for luck.

J. A. Lindon, a British writer of humorous verse, sent me four specimens of unintended verse which he said he had found in a *Manual of Practical Anatomy,* by someone named Cunningham:

Scrape the fatty tissue
From the popliteal surface
Of the femur with the handle of a knife

Put the grove in deep
And its margins meet,
And the vessels are therefore enclosed in the gland.

It is bounded on the left
By a deep and narrow vertical cleft.

For it has to cross the median plane
To reach the beginning of the portal vein.

Lindon also sent the following items, with apologies for not having the first names of authors:

When parallel rays
Come contrary ways
And fall upon opposite sides.
Dr. Smith's *Optics.*

. . . . is called into play to steady the wrist
When the hand grasps the object or makes a fist.
—Grant, A Method of Anatomy

Herman M. Frankel passed along this gem from Werner Heisenberg's *Physics and Philosophy:*

Every word or concept
Clear as it may seem to be,
Has only a limited range
Of applicability.

Arthur Koestler, in *Roots of Coincidence,* quotes a passage from another physicist that can be read as follows:

Particles of
Imaginary mass,
Interacting together
Like a frictionless gas.

On October 21, 1979 (my birthday, incidentally), the *New York Times* published my letter:

Until I read your account of the Florida voting results (Oct. 15), I hadn't known that Jody Powell likes to speak in rhymed doggerel. Here is his comment, no word or punctuation mark altered:

They put in
The best they had
And we put in
The best we had,
And we beat them
And beat them bad.

To which I add: It makes me sad, to see the ways, we're being had.

Eugene McCarthy, one-time candidate for president, was quoted in the *New York Times* (November 7, 1968) as remarking, with reference to Richard Nixon, "Under the shadow of his wings, we can think of other things."

John Leonard ended his "Critic's Notebook" column in the *New York Times* (August 7, 1980) with this quatrain:

After a leisurely lunch, he forgot to look
Both ways while crossing the rue.
There may be a laundry truck
With your name on it, too.

Marvin Minsky, in his *Society of Mind,* item 3.5, wrote: "A real child can go to bed—yet still build towers in its head."

Sheldon Glashow, the physicist, in Chapter 14 of his autobiography *Interactions,* unintentionally warbles:

They will tell you
Why copper is red,
Why the sky is blue,
How a candle burns,
And what makes dew.

Hans Moravec, in his book on artificial intelligence, *Mind Children,* dedicates the book this way:

In memory of
My father, who taught me to tinker,
To my mother who taught me to read,
To Ella who made me complete.

Beneath this I scribbled, "And to software that gave me a creed."

William Henry, in a *Time* cover story about gossip (March 5, 1990) ended a paragraph with:

The rockies may tumble,
Gibralter may crumble,
They're only made of clay.
But gossip is heaven sent
And here to stay.

As a boy I recall my father coming home from his office one afternoon and telling us that he had been to the post office with his accountant to check the contents of the mail box. "Anything there?" dad asked. The accountant replied with an unintended jingle in the same meter as the last line of "Punch, Brother, Punch": "Nothing but a notice that the box rent's due."

My wife Charlotte spotted the next two items:

The stars start to fade,
And soon the sun is out,
Baking the corn and soybeans
And anyone not in the shade.
—from a *New York Times* (August 18, 1972) story about midwest farms.

Neither snow nor rain, nor heat,
Nor gloom of night,
Stops the mail from getting through,
But potholes might.
—from an AP dispatch (May 20, 1986) about potholes in Windsor Heights, West Virginia.

Brian Agran found a bit of unintended doggerel at the end of the first paragraph the previous chapter:

Some accidental patterns
Involving the numbers pi and e,
And then some whimsical instances of
Accidental poetry.

In the category of unintended meanings that turn serious poetry into farce, the classic example occurs near the end of Robert Browning's *Pippa Passes:*

But at night, brother howlet,
over the woods
Toll the world to thy chantry;
Sing to the bats' sleek
sisterhoods
Full complines with gallantry:
Then, owls and bats,
Cowls and twats,
Monks and nuns, in a cloister's
moods,
Adjourn to the oak-stump pantry!

Twat, then pronounced with a flat *a,* has never had any meaning except the one it has now, but when Browning came across it in a book he took it to be the name of something worn by nuns. The fact that he never removed it from his famous poem suggests that no one had the courage to explain the mistake to him!

Other lines in Browning's poetry are almost as unintentionally funny. In "The Flight of the Duchess," for instance, he speaks appropriately of Paris as "The Land of Lays."

Here are the first two stanzas of Emily Dickinson's unfinished Poem 566, in *The Complete Poems of Emily Dickinson:*

A Dying Tiger – moaned for Drink –
I hunted all the Sand –
I caught the Dripping of a Rock
And bore it in my Hand –

His Mighty Balls – in death were thick –
But searching – I could see
A Vision on the Retina
Of Water – and of me –

I was stunned for a moment before I realized that Emily was referring to the tiger's mighty eyeballs. Her well known line, "There is no frigate like a book," was called to my attention by the poet George Starbuck, whose elegant poems often swarm with clever word play.

A surprising clerihew, reprinted with adorning sketches by Gilbert Chesterton, is on page 49 of *The First Clerihews* (Oxford Press, 1982), by E. Clerihew Bentley and his friend Chesterton, who surely did not realize how it could be misinterpreted:

Mr. Oscar Wilde
Got extremely riled.

He ejaculated, "Blow me
If I don't write 'Salomé.' "

A line in Milton's *Paradise Lost* (Book 4) refers to Adam's "fair large front." Henry Christ, who teaches high school literature in Florida, tells me that whenever he reads from Alfred Noyes's "The Highwayman" the line, "One kiss, my bonny sweetheart, I'm after a prize tonight," his students break into giggles.

"McAndrews's Hymn," by Rudyard Kipling, contains these startling lines:

'Twas on me like a thunderclap—it racked me through an' through—
Temptation past the show o' speech, unnameable an' new—
The Sin against the Holy Ghost? . . . An' under all, our screw.

Verbatim, a British journal devoted to linguistics, ran an article in its Winter, 1989, issue titled "Red Pants," by Robert Sebastian. It is devoted entirely to unintended meanings in prose and poetry. I will cite only some verse specimens.

Francis Thompson, explaining how "day's dying dragon" colors a sunset sky, in his poem "A Corymbus for Autumn," wrote: "panting red pants into the west."

"In this trouser category," Sebastian writes, "Thompson must share the limelight with Coleridge" who wrote in "Kubla Khan," "As if this earth in fast thick pants were breathing." And Shelley, in "Epipsychidion," tells how

. . . . the slow, silent night
Is measured by the pants of their calm deep.

Closely related to pants are some lines in Milton's *Paradise Lost* (Book 1, lines 236–7) which tell how Mount Aetna will thunder:

And leave a singed bottom all involved
With stench and smoke.

Today's pejorative meaning of *pansy* tends to ruin such lines as "With rue and the beautiful Puritan pansies" (Poe, in "For Annie"). A current meaning of *gay* similarly spoils such lines as "Why is my neighbor's wife so gay?" (Chaucer, *Prologue of the Wife of Bath*); "If nature made you so grateful, don't get gay" (William Vaughn Moody, "The Menagerie");

"Never lacked gold, and yet went never gay" (Shakespeare, *Othello,* Act II, Scene 1); and "They know that Hamlet and Lear are gay" (William Butler Yeats, "Lapis Lazuli").

The reader can look at Sebastian's article for even funnier examples from classic prose. For unintended pornography in the Sherlock Holmes canon, see John Bennett Shaw's celebrated lecture "To Shelve or to Censor," available, alas, only in the obscure journal *Shades of Sherlock,* Volume 5, No. 2, August, 1941, pp. 4–12.

In "The Old Stage Queen" Ella Wheeler Wilcox unintentionally has her queen hit the floor in the following stanza:

She rises to go. Has the night turned colder?
The new Queen answers to call and shout;
And the old Queen looks back over her shoulder,
Then all unnoticed she passes out.

The first publication of Thomas Hardy's poem "The Caged Goldfinch," contained a stanza telling how the cage and bird came to be placed on a woman's grave:

True, a woman was found drowned the day ensuing,
And some at times averred
The grave to be her false one's, who when wooing
Gave her the bird.

Hardy thought it best to omit the stanza when someone told him about the unintended meaning of its last line.

20

Packing Squares

"It *is* a Square! . . . Beautiful! Beau-ti-ful! Equilateral! *And* rectangular!"

—LEWIS CARROLL,
A Tangled Tale, Knot Two

In Chapter 10 I discussed the unsolved general problem of packing n identical circles into squares of minimum area so that the circles do not overlap. Packing n identical squares into larger squares of minimum area presents a similar problem, which except for very low values of n is also unsolved. After summarizing the sparse results pertaining to this difficult task, I shall turn to some other curious questions concerning the packing of squares, then make a brief excursion into the third dimension.

Like the task of packing circles into a square, the task of packing squares into a square can be viewed in two ways. On the one hand, if the outer square is assumed to have a side of length 1, the problem is to determine how large n identical squares can be and still fit into the outer square. On the other hand, if the identical squares are assumed to have a side of 1 (in which case they are called unit squares), the problem is to determine the smallest square into which the n unit squares will fit. The latter procedure is the one that will be considered here. Amazingly, minimal solutions have been found only in those cases where n is the square of an integer or is equal to 2, 3 or 5.

Call the side of the square to be packed with unit squares k. Figure 119 displays packings for the best, or lowest, values known for k when the number of unit squares n ranges from 1 through 15, with brief comments on each packing. Obviously whenever n is a square number, k is the square root of n, and in the cases where n equals 2, 3 or 5 it is not difficult to prove that the values of k shown in the illustration are minimal. When n is not a square number, the side k must be greater than the square root of n, and it can always be taken to be less than or equal to the lowest integer greater than the square root of n.

When n is not large and equals $a^2 - a$ for some integer a, it is conjectured that the side k of the minimum bounding square is equal to a. Or to put the conjecture geometrically, the unit squares in an $(a - 1)$-by-a array cannot be rearranged to fit into a square of side smaller than a. If the conjecture is true, it follows at once that a square of the same size is also minimal when the number of unit squares is increased up to and including a^2. In order to see this, assume that the bounding square can have a side smaller than a and still accommodate one or more unit squares in the empty, or ath, row. Removing the extra square or squares will leave $a^2 - a$ squares inside a square of side smaller than a, which contradicts the conjecture.

The smallest value of n for which the conjecture does not hold is not known. Ronald L. Graham of Bell Laboratories has found that the conjecture is violated when n equals $40^2 - 40$, or 1,560 (that is, when the outer square measures 40 on a side), but he believes the smallest value of n that violates the conjecture is considerably lower, perhaps closer to $10^2 - 10$, or 90. The smallest n for which the best packing requires a tilting of squares at any angle other than 45 degrees is also not known.

These results, as well as those given in Figure 119, are taken from the only reference I know for this fascinating problem, the paper "Geometrical Packing and Covering Prob-

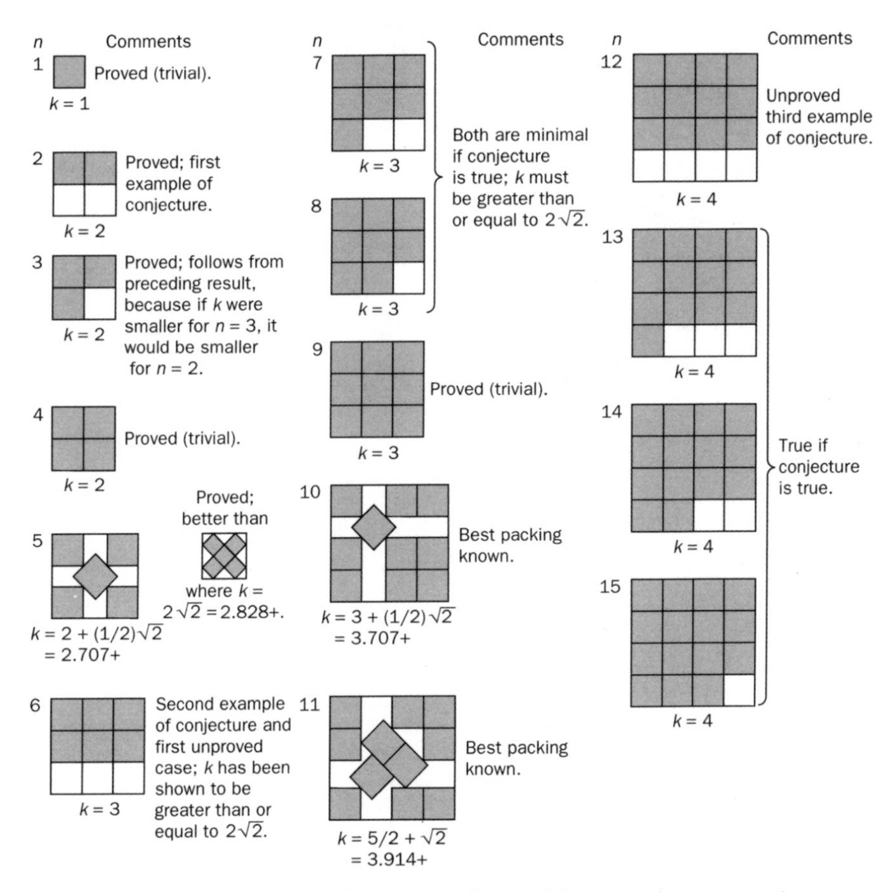

FIGURE 119 Best results known for packing n unit squares into the smallest square possible when n ranges from 1 through 15

lems," by the Dutch mathematician F. Göbel. When n equals 16, of course, the value of k is 4. Packings for the best values known for k when n equals 17 and 18 are shown in Figure 120. If the conjecture mentioned above is true, k equals 5 when n equals 20 through 25.

The case $n = 19$ is of special interest. Early in 1979, after reading about the packing of unit circles into squares, Charles F. Cottingham began to investigate the packing of unit squares into squares. Unaware of previous work in this area, Cottingham duplicated Göbel's results up through $n = 25$, with one exception. For the case $n = 19$ he obtained a slightly better

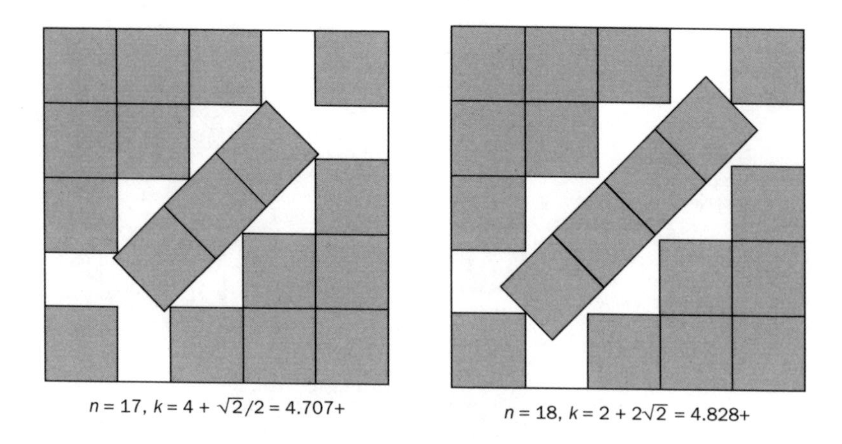

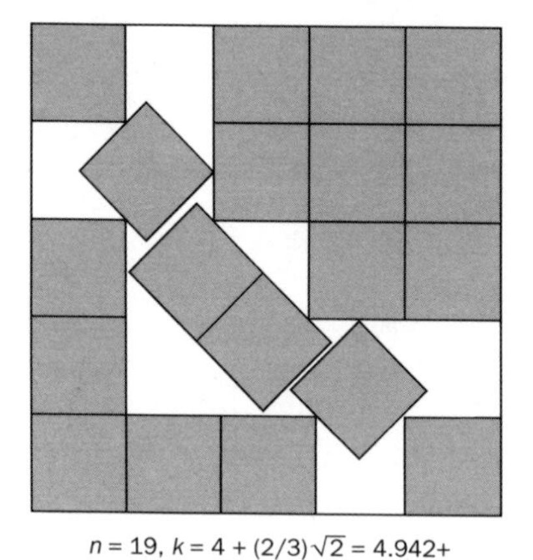

FIGURE 120 Packings for $n = 17$, $n = 18$ and $n = 19$

packing than the one shown at the bottom of Figure 120. Can you find it? Cottingham also improved on Göbel's patterns for certain values of n higher than 25.

As n becomes larger the task of proving that a packing square is minimal (when n is not a square number) becomes increasingly difficult. Even a proof for the case $n = 5$ is not trivial. One way to go about determining the minimum side length of the square that bounds five unit squares is to divide it into four equal square regions. Since five unit squares must go inside, the pigeonhole principle ensures that the centers of at least two of the unit squares must be inside or on the border of at least one of the four square regions. The final step is to show that a square region large enough to accommodate the centers of two unit squares cannot be smaller than $1+(1/4)\sqrt{2}$ on a side. Hence the side of the minimal bounding square must be twice that length, or $2+(1/2)\sqrt{2}$.

No general procedure for finding minimum bounding squares is anywhere in sight, but in 1975 Paul Erdös and Graham published a paper in which they proved a remarkable theorem. They showed that as the number of unit squares increases to a sufficiently large number, clever packings can lower the amount of wasted space to an area that never exceeds $k^{7/11}$, or $k^{.636+}$, where k is the side of the bounding square. (The area of wasted space is zero when there is a square number of unit squares.)

Hugh Montgomery later lowered this asymptotic bound slightly, to $k^{(3-\sqrt{3})/2}$, or $k^{.633+}$, and it can probably be lowered further. Graham conjectures that as n approaches infinity the ultimate bound may well be $k^{.5}$, or the square root of k. In a paper published in 1978 Klaus F. Roth and Robert C. Vaughan, two British mathematicians, showed that the bound cannot have a limit below $k^{.5}$. Therefore as the matter now stands, with respect to optimal packings of large numbers of unit squares into minimal bounding squares, the wasted area will be greater than or equal to $k^{.5}$ and less than or equal to $k^{.633+}$.

To dramatize how much space can be saved when the number of unit squares is large, Graham considers the packing of unit squares into a large square of side 100,000.1. If the squares are conventionally packed in straight rows, $100{,}000^2$ unit squares can be fitted in. This packing leaves an extremely thin, linear border of empty space into which no unit square can fit along two sides of the outer square. If the squares are tilted and packed according to the Erdös-Graham-Montgomery technique, however, more than 6,400 additional unit squares can be fitted in!

Now consider the sequence of squares whose sides are the consecutive integers 1, 2, 3, 4, 5. . . . The consecutive areas of these squares form the infinite sequence 1, 4, 9, 16, 25. . . . , and the consecutive partial sums of this sequence are 1, 5, 14, 30, 55. . . . Does the sequence of sums include a square number? Yes, but only one: 4,900. In the mid-1960's this number, which is the square of 70 and the sum of the first 24 square numbers, suggested the following problem: Can the 70-by-70 square be packed with the first 24 squares whose sides are consecutive integers? The answer to the question is no, but there are packings of squares from the set that omit the 7-by-7 square and leave as little as 49 unit squares of wasted space. This packing is thought to be minimal, although I know of no proof of the conjecture. Nor do I know whether a packing with less wasted space can be achieved if the order-70 square is viewed as a cylindrical or toroidal surface.

Will a set of these consecutive squares (starting with 1 or a higher number) tile a rectangle? This question, first asked by Solomon W. Golomb, remains unanswered. If the answer is no, as is suspected, Golomb suggests the following refinement: What pattern of consecutive squares starting with 1 can be packed into a larger square to leave the smallest percentage of wasted space? The same question can be asked of a bounding rectangle. I know of no work that has been done so far on any of these questions.

No square can be packed with squares of sides 1, 2, 3, 4, 5 . . . , and the same is probably true for a rectangle, but is it possible to tile the infinite plane with consecutive squares starting with 1? This unusual problem was posed by Golomb in 1975. Consecutive squares of sides 1 through 11 can be spiraled around a center as shown in Figure 121, but that tiling leaves an unavoidable hole: the space marked *X*. No algorithm for tiling the entire plane with consecutive squares beginning with any integer has been found, but the task has not been proved impossible.

Figure 122 shows one way consecutive squares can be placed to form a simply connected region (an area with no holes) that will cover more than three-quarters of the plane. The southeast quarter is covered by squares with sides in the familiar Fibonacci sequence: 1, 2, 3, 5, 8, 13. . . . The southwest quarter is covered by a truncated Lucas sequence that begins 7, 11, 18, 29 . . . and a thin strip of overlap from the squares on the right. (The Lucas sequence, which begins 1, 3, is the simplest of the generalized Fibonacci sequences.) The northeast quarter is covered by squares in a truncated Fibonacci sequence that begins 6, 9, 15, 24 . . . and a thin strip of overlap from the squares below. Verner E. Hoggatt, Jr., then editor of *The Fibonacci Quarterly,* has shown that no numbers are duplicated in these three sequences. The squares with the sides that were omitted from the sequences are arranged along the top of the Lucas squares. The black spot in

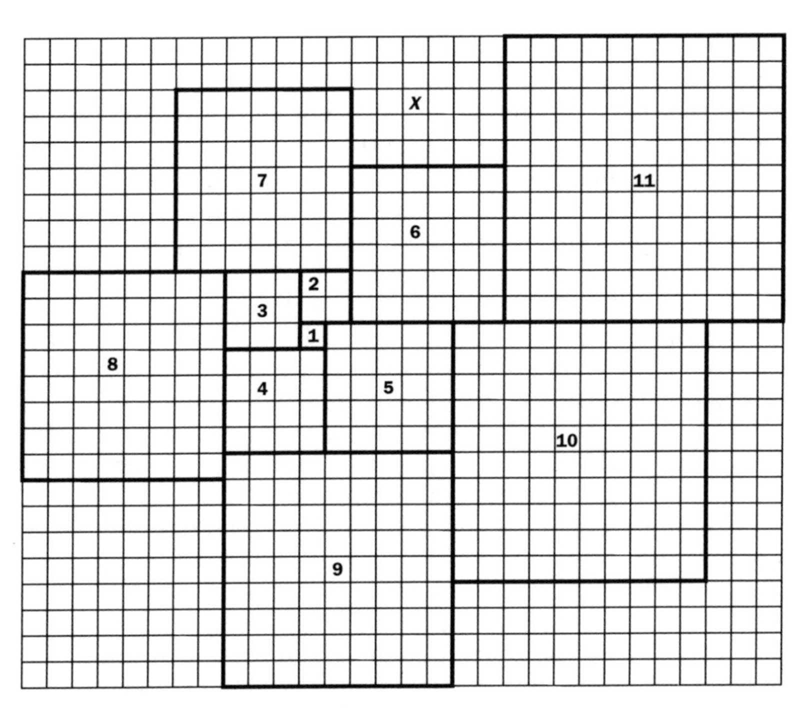

FIGURE 121 Can the plane be tiled with consecutive squares?

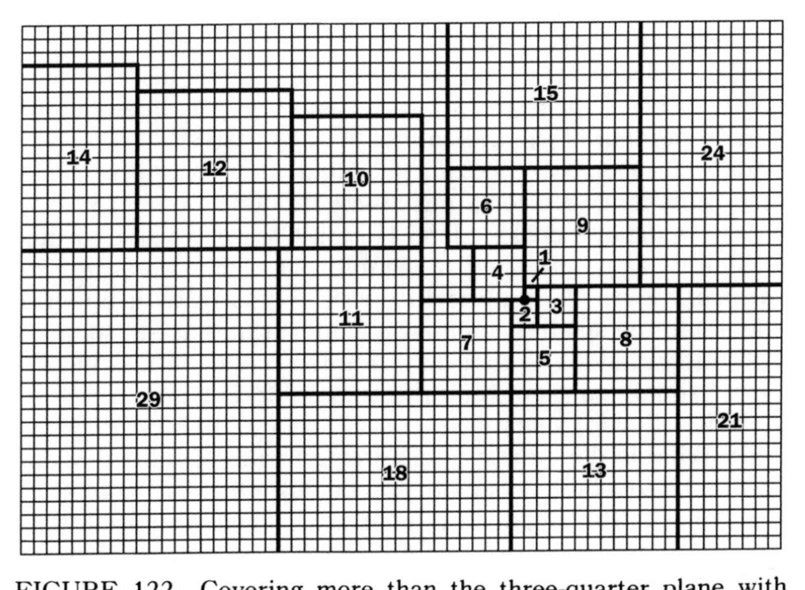

FIGURE 122 Covering more than the three-quarter plane with consecutive squares

the illustration is the origin point of the Cartesian plane. This pattern of consecutive squares covers more than three-quarters of the plane, of course, but it is not known whether even the quarter plane can be tiled exactly with consecutive squares.

There are many ways to tile the entire plane with squares whose sides are not consecutive but only different. Here is one way, which makes use of the fact that a square can be dissected into as few as 21 different squares (see Figure 77). The smallest square in any such dissection is necessarily an interior one, and so to tile the plane let a dissected square, with its 21 subsquares, serve as the smallest square in a larger replica of the same pattern. The larger (tiled) square can then be the smallest square in a still larger replica, and so on. By continuing in this way it is possible to completely cover the infinite plane with squares of different sizes.

For more than a decade it has been known that any set of squares with a total area of 1 (the sides need not be rational) can be packed without overlap into a square of area 2 (see my *Mathematical Carnival,* Chapter 11). Suppose a rectangle has unit width. What is the minimum length that will allow the rectangle to bound any set of squares with a combined area of 1? A long-conjectured answer of $\sqrt{3}$ was verified by Daniel J. Kleitman and Michael M. Krieger in 1970 (see the bibliography). In a 1975 paper the same two authors showed that the rectangle of smallest area into which any set of squares with a total area of 1 can be packed has sides of $2/\sqrt{3}$ and $\sqrt{2}$. In each of these results the wasted area is huge—more than 60 percent of the area of the set of squares.

A classic square-packing problem known as Mrs. Perkins' Quilt is named for a puzzle mentioned in one of Henry Ernest Dudeney's books. (John Horton Conway's work on this problem is also described in Chapter 11 of my *Mathematical Carnival.*) Here the task is to divide a square with integral side n into the smallest number of nonoverlapping subsquares with integral sides so that no space is wasted. In this puzzle the subsquares are not required to be different.

The following question is closely related to Mrs. Perkins' Quilt: What is the largest number of subsquares (allowing duplicates) into which no square can be cut? Or to put it another way, what is the smallest number n of subsquares such that a square can always be cut into n and all higher numbers of subsquares? It is not hard to show that a square cannot be cut into two, three or five subsquares, but it can be cut into six subsquares and all higher numbers of subsquares. (To cut a square into six subsquares, divide a 3-by-3 square into nine unit squares and then mark off one subsquare of side 2 and five of side 1.) Hence the answer to the question phrased the first way is 5, and the answer to the question phrased the second way is 6.

The natural extension of this problem to cubes turned out to be considerably more difficult. It was established that a cube can be subdivided into n subcubes when n equals 1, 8, 15, 20, 22, 27, 29, 34, 36, 38, 39, 41, 43, 45, 46, 48, 49, 50, 51, 52, 53, 55 and all higher numbers. The task was proved to be impossible for all the remaining values of n except one: $n = 54$. For many years the question of whether a cube could be cut into 54 subcubes remained a perplexing problem.

As long as the case $n = 54$ was unsolved it was impossible to answer the following question: What is the largest number of subcubes, not necessarily different, into which a cube cannot be cut? The question became known as the Hadwiger problem because it had been framed in 1946 by a Swiss mathematician, Hugo Hadwiger of Bern. The answer was thought to be 54, but a more recent dissection of a cube into 54 subcubes was independently found by two other residents of Bern, Doris Rychener, a flute teacher, and A. Zbinden of the International Business Machines Corporation. The solution was published by Richard K. Guy in 1977. Figure 123, based on an illustration provided by Guy, shows 42 cubes of side 1, four of side 2, two of side 3 and six of side 4. The four pieces

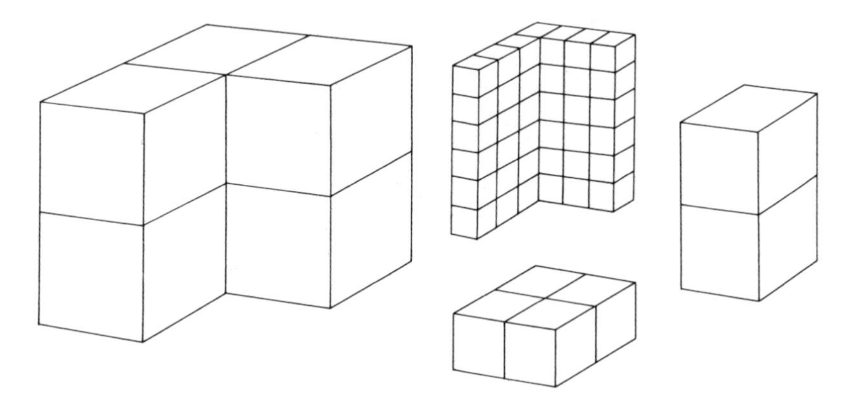

FIGURE 123 Doris Rychener's dissection of the 8-by-8-by-8 cube into 54 subcubes

shown go together in an obvious way to make a cube of side 8.

It is now possible to answer Hadwiger's question. The largest number of cubes into which a cube cannot be cut is 47; or 48 is the smallest number such that a cube can be cut into that number of subcubes and all higher numbers of them. Incidentally, it is not possible to cut a cube into any number of subcubes of different sizes. The proof, one of the most beautiful in combinatorial geometry, can be found in my addendum to William T. Tutte's article "Squaring the Square," reprinted in my *2nd Scientific American Book of Mathematical Puzzles & Diversions.*

Many curious square-packing problems concern only a single square that fits into a region of a specified shape. For example, what is the largest square that can be inscribed in a regular pentagon of side 1?

ANSWERS

Figure 124 shows the best known packing (as of 1979) of nineteen unit cubes into a square.

The problem of finding the largest square that will go inside a regular pentagon was posed by Fitch Cheney in the *Journal of Recreational Mathematics,* and answered by him in

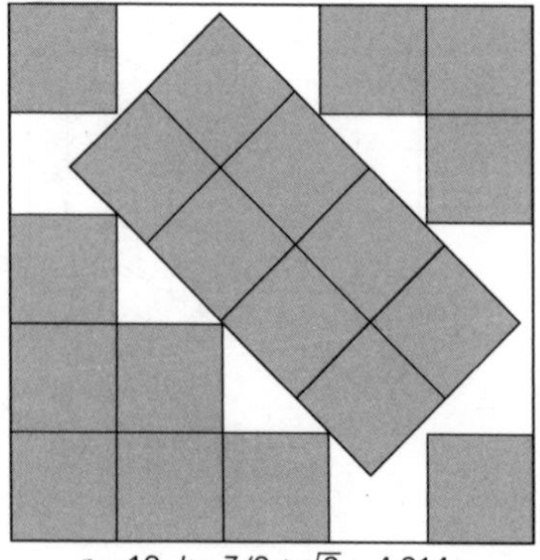

$n = 19$, $k = 7/2 + \sqrt{2} = 4.914+$

FIGURE 124 Solution to the packing of 19 unit cubes

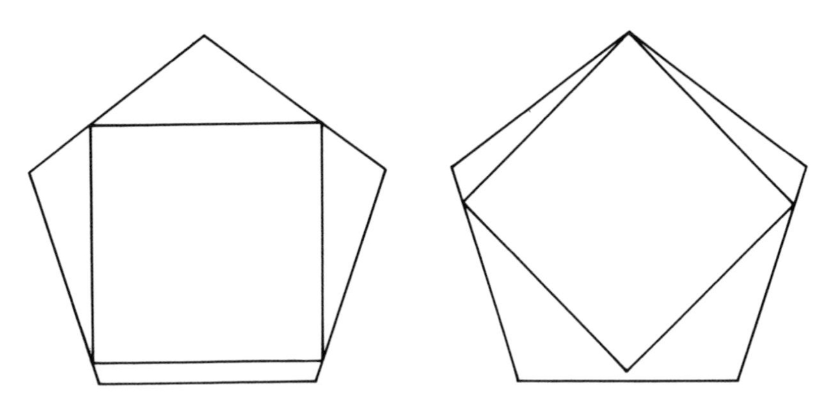

FIGURE 125 Solution to packing the largest square in a regular pentagon

a 1970 issue (see the bibliography). It is easy to suppose the square shown at the left in Figure 125 is the largest because the slightest tilting of the square moves one of its corners outside the pentagon. The correct answer, however, is shown at the right. Assuming that the pentagon has a side of 1, the square at the left has a side of 1.0605+, whereas the square at the right has a side of 1.0673+. Note that the bottom corner of the square at the right does not quite touch the base of the pentagon.

ADDENDUM

In 1979, Walter Trump, of Nürnberg, West Germany, sent me the pattern for packing 11 unit squares shown in Figure 126. It lowers k to 3.877+. Walter R. Stromquist, of Paoli, Pennsylvania was able to show that this answers a question raised by Ronald L. Graham: What is the smallest number of unit squares for which the densest packing into a square requires a tilting of the squares at an angle other than 45 degrees? In this case the angle is 40.18 degrees. The pattern was later found by others, and as far as I know is the densest known packing for $n=11$.

In a 1984 letter Stromquist improved Göbel's results for $n=18$ and $n=26$. The two patterns are shown in Figure 127.

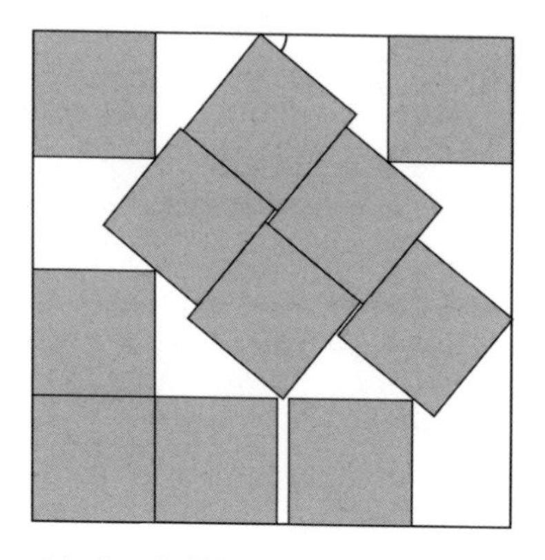

.‹E 126 $n=11$, $k=3.877+$, the tilt angle$=40.18°$

On the left the tilt angle is 4.82+ degrees, on the right it is 27.58+ degrees. The $n=18$ result had earlier been found by Pertti Hämäläinen, of Finland, who sent it to me in 1980, and by Mats Gustafson, of Ludvika, Sweden, who provided it the following year.

Robert T. Wainwright was the first reader to find an improved packing for 19 unit squares. His solution, shown in Figure 128, lowers k to $(4\sqrt{2/3})+3=4.885+$, the current record. The pattern was also discovered by Robert Ammann, Joseph Crowther, Paul Engler, Gerald Gough, Hugh Everett, David Hobby, Richard Holzsayer, David Kitchens, Milos Konopasek, Evert Stenlund, and Douglas Stoll.

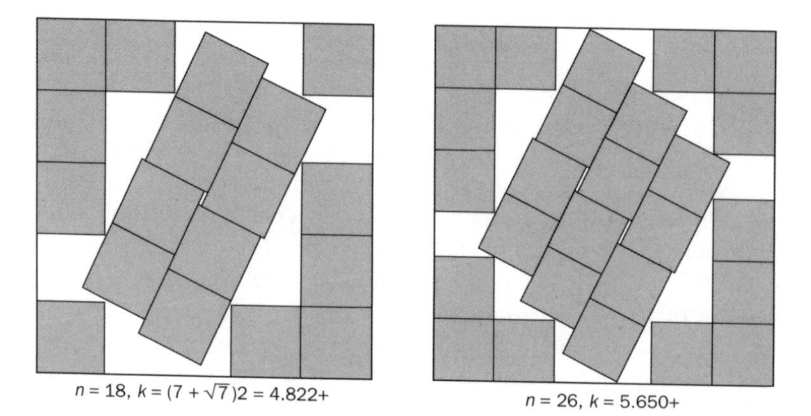

FIGURE 127 $n=18$, $k=(7+\sqrt{7})2=4.822+$ $n=26$, $k=5.650+$
Improved packings by Walter Stromquist

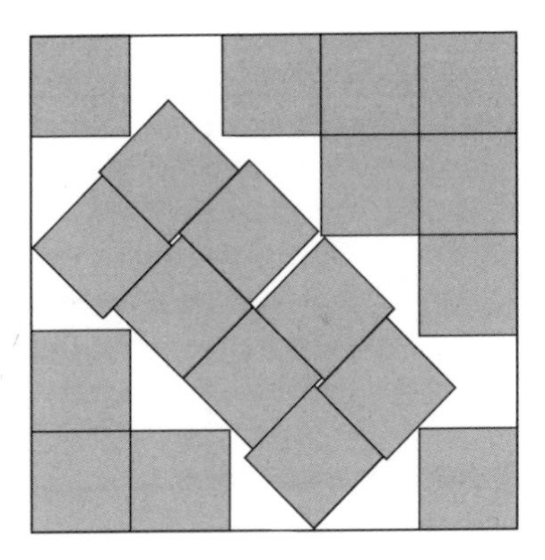

FIGURE 128 Packing of 19 unit squares into a square of side $3+(4/3)\sqrt{2}$, or 4.885+

I mentioned the old problem of tiling the 70×70 square with squares taken from the set of squares with sides 1 through 24. A perfect tiling with all 24 squares is impossible, and the densest known tiling omits the 7^2 for a covering of all but 49 square units. I asked if this could be improved if the 70×70 square had its sides joined to make a torus or a cylinder. The answer is yes. The best solution for the cylinder (see Figure 129) came from Robert Reid. It omits the 2^2 and 6^2 leaving an uncovered area of 40 square units. Frank L. Paulsen sent the best packing for the torus (see Figure 130). It omits the 3^2 and 5^2, leaving an uncovered area of 34 square units. I will be pleased to hear of any improvements on either task.

Robert Wainwright, intrigued by the 70×70 problem, thought of the following related task. Partial sums in the sequence $1(1^2)+2(2^2)+3(3^2)$ . . . are square numbers. What is the smallest square, he asked, in which a set of such consecutive squares can be perfectly packed (no overlap or holes). In other words, we wish to perfectly pack a square with one unit square, two squares of side 2, three of side 3, and so on. It is not hard to show that if n is the side of the enclosing square, it cannot have values 1 through 5. Wainwright could not find solutions for $n=6$ through 11, but was unable to prove impossibility. He conjectures that the $n=12$ square shown in Figure 131 is the smallest that can be perfectly packed. Can anyone do better?

I gave a way to tile the plane with distinct integer-sided squares based on the dissection of a square into 21 unequal

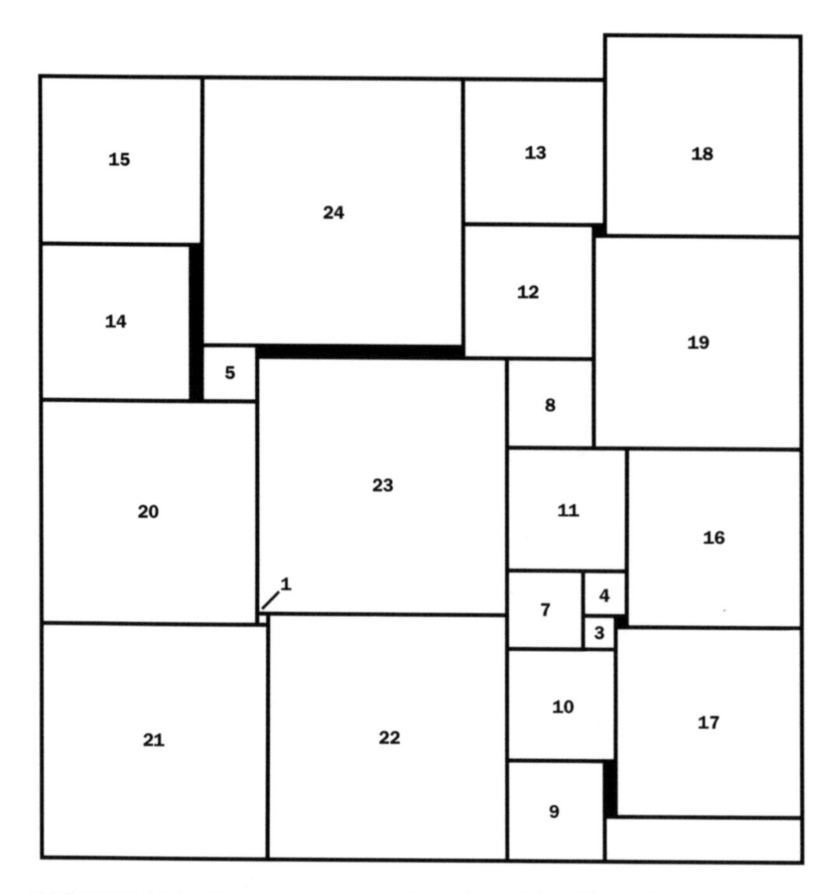

FIGURE 129 Best known solution of the 70×70 problem on a cylinder

squares. Another way to do it is to start with a unit square so dissected, then whirl around it squares of 1, 2, 3, 5, 8, 13. . . . in the Fibonacci sequence. It has been shown that the number of essentially different ways of tiling the plane with unequal integer squares is 2 to the power of aleph-null.

In 1964 D. E. Daykin asked if space could be tiled with unequal integer cubes. In 1982 Robert J. M. Dawson (see the bibliography) proved this to be impossible, and extended the result to all higher Euclidean spaces. In 1988 Dawson broadened this result by removing the integer restriction and showing that no space tiling is possible with unequal cubes even when the sides can be real numbers.

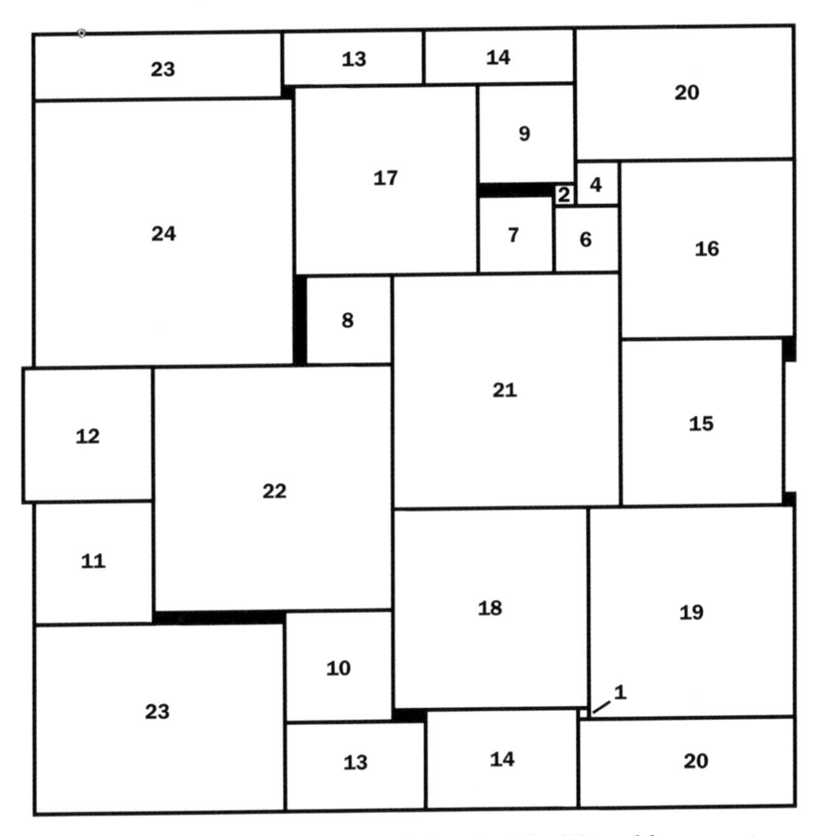

FIGURE 130 Best known result for the 70 × 70 problem on a torus

Golomb's problem of tiling the plane with consecutive squares, starting with 1, remains unsolved. However, Brian Astle, of Princeton, New Jersey, wrote to explain how the plane could be tiled with consecutive squares in such a way as to leave an arbitrarily small fraction of the plane untiled. Every tile in the sequence is at a finite distance from the origin point. Astle's method, too complicated to explain here, remains unpublished. Golomb tells me that he may have found an algorithm for solving his problem, but there are still gaps in the proof that he has yet to fill.

It is easy to see that it is not possible to tile the plane with unit squares without having an infinite number of squares that share an edge with an adjacent square. Can unit cubes be close-packed to fill space without any pair of cubes meet-

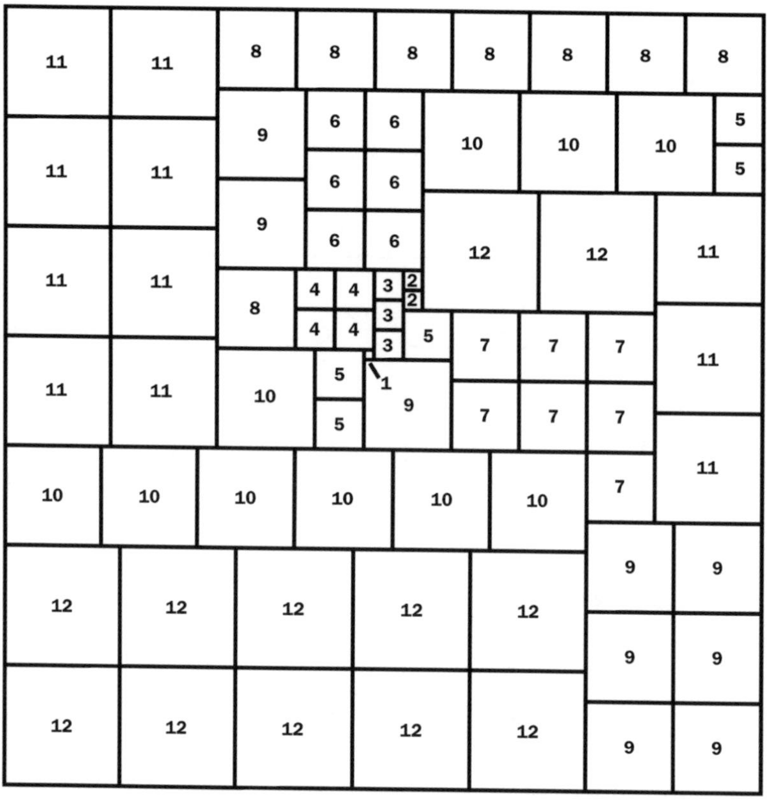

FIGURE 131 Is this the smallest square that solves Wainright's problem?

ing face to face? Surprisingly, the answer is yes, though it is difficult to visualize or draw pictures of such packing. Raphael M. Robinson was the first to discover such a tiling, which he describes in a paper published in *The Mathematical Gardner* (see the bibliography). After Robinson's article was written, Basil Gordon, at the University of California, Los Angeles, found a much simpler tiling, and proved that such tilings exist for hypercubes in all higher dimensions. Robinson reported this in a postscript, but Gordon has not yet published his proof.

BIBLIOGRAPHY

"Problem E724." Hugo Hadwiger, in the *American Mathematical Monthly,* 53, 1946, page 271. William Scott, commenting on the problem in Vol. 54, 1947, pages 41–42, proved that a cube can be cut into any number of smaller cubes of 54 or greater. He generalized the problem to higher Euclidean spaces, giving bounds for hypercubes in four, five, and six dimensions.

"Inscribable Square." Fitch Cheney, in *Journal of Recreational Mathematics,* 3, 1970, pages 232–233.

"Packing Squares in Rectangles." Daniel J. Kleitman and Michael M. Krieger, in *Annals of the New York Academy of Sciences,* 175, 1970, pages 253–262.

"Decomposition of a Cube Into Smaller Cubes." Christopher Meier, in the *American Mathematical Monthly,* 81, 1974, pages 630–633.

"On Packing Squares with Equal Squares." Paul Erdös and Ronald L. Graham, in *Journal of Combinatorial Theory,* Series A, 19, 1975, pages 119–123.

"An Optimal Bound for Two Dimensional Bin Packing." Daniel J. Kleitman and Michael M. Krieger, in the *16th Annual Symposium on Foundations of Computer Science,* IEEE Computer Society, Institute of Electrical and Electronics Engineers, 1975.

"The Heterogeneous Tiling," Problem 388. Solomon W. Golomb, in *Journal of Recreational Mathematics,* 8, 1975–76, pages 138–139.

"Research Problems." Richard K. Guy, in the *American Mathematical Monthly,* 84, 1977, page 810.

"Inefficiency in Packing Squares with Unit Squares." L. F. Roth and R. C. Vaughan, in *Journal of Combinatorial Theory,* Series A, 24, 1978, pages 170–186.

"Geometrical Packing and Covering Problems." F. Göbel, in *Packing and Covering in Combinatorics,* edited by A. Schrijver. Amsterdam: Tweede Boerhaavestraat, 1979.

"Can Cubes Avoid Meeting Face to Face?" Raphael Robinson, in *The Mathematical Gardner,* edited by David A. Klarner. Wadsworth, 1981.

"On Filling Space with Different Integer Cubes." Robert J. Dawson, in *Journal of Combinatorial Theory,* Series A, 36, 1984, pages 221–229.

"On the Impossibility of Packing Space with Different Cubes." Robert J. Dawson, in *Journal of Combinatorial Theory,* Series A, 48, 1988, pages 174–188.

21

Chaitin's Omega

A fascinating unpublished paper titled "On Random and Hard-to-Describe Numbers" has been written by Charles H. Bennett, a mathematical physicist at the Thomas J. Watson Research Center of the International Business Machines Corporation. It begins by recalling certain paradoxes involving integers, in particular a paradox that seems to stand in the way of calling any specific integer a random one. The resolution of the paradox makes randomness a property that most integers have but cannot be proved to have, a subject that was treated by Gregory J. Chaitin, also of the Watson Research Center in a 1975 *Scientific American* article. Bennett goes on to consider the senses in which irrational numbers such as pi can be said to be random. And he discusses at length the irrational number Ω, recently discovered by Chaitin, which is so random that in the long run no gambling scheme for placing bets on its successive digits could do better than break even.

The number Ω has other unusual properties. To begin with, it can be defined precisely but it cannot be computed. Most remarkable, if the first few thousand digits of the number were known, they would, at least in principle, provide a way to answer most of the interesting open questions in mathematics, in particular those propositions that if false could be refuted in a finite number of steps. The following discussion of these issues is taken from Bennett's paper, which begins with a simple variant of Berry's paradox. Named after G. G. Berry, the Oxford librarian who discovered it, the paradox was first published by Bertrand Russell and Alfred North Whitehead in *Principia Mathematica.*

By Charles H. Bennett

The number "one million, one hundred one thousand, one hundred twenty one" is unusual in that it is, or appears to be, the number named by the expression: "The first number not nameable in under ten words." This expression has only nine words, however, and so there is an inconsistency in regarding it as a name for 1,101,121 or any other number. Berry's paradox shows that the concept of nameability is inherently ambiguous and too powerful to be used without restriction. In a note in *The American Mathematical Monthly* [April, 1945] Edwin F. Beckenbach pointed out that a similar paradox arises when one attempts to classify numbers as either interesting or dull: There can be no dull numbers, because if there were, the first of them would be interesting on account of its dullness.

Berry's paradox can be avoided and even tamed, however, if the definition of nameability is restricted, that is, if an integer is considered to be named when it has been calculated as the output of a computer program. To standardize the notion of computation in this definition a simple idealized computer known as a universal Turing machine is introduced. This machine will accept a program in the form of a sequence of 0's and 1's on an input tape and write the results of its computation, again in the form of a sequence of binary digits, on an output tape at the end of the computation. A third tape is used during the computation to store intermediate results. (The use of separate tapes for input, output and memory is a conceptual convenience rather than a necessity; in the earliest Turing machines the same tape served all three purposes.) As is well known, a universal Turing machine can do anything the most powerful digital computer can do, although considerably slower. More generally, it can perform even the most complicated manipulation of numerical or symbolic information as long as the manipulation can be expressed as a finite sequence of simple steps in which each step follows from the preceding one in a purely mechanical manner without the intervention of judgment or chance.

An integer x can now be named by specifying a binary sequence p that, when it is given to the Turing machine as input, causes the machine to calculate x as its sole output and then halt. There can be no doubt that the program p does indeed describe x. Hence the universal Turing machine provides an unambiguous but flexible language in which a number can be described according to any of the ways it might be effectively calculated. (For example, the number 523 might be described as the 99th prime number, as $(13 \times 41) - 10$ or, more directly, as the binary sequence 1000001011.) Every integer is nameable in this language, because even an integer with no distinguishing properties can always be described by simply giving its binary sequence.

Returning to the question of interesting and dull numbers, an interesting number can now be defined, without paradox, as one computable by a program with considerably fewer bits, or binary digits, than the number itself. This short description would reflect some special feature by which the number could be distinguished from the general run of numbers. By this definition, then, $2^{65,536} + 1$, the first million digits of pi and (17!)! are interesting numbers. [The exclamation point is the factorial sign: $n!$ equals $1 \times 2 \times 3 \times \ldots n$.] Conversely, a dull, or random, number is one that cannot be significantly compressed, that is, one whose shortest description has about as many bits as the number itself. This algorithmic definition of randomness as incompressibility, which is reviewed in the article by Chaitin mentioned above, was developed in the 1960's by several mathematicians, including Ray J. Solomonoff and Chaitin in the U.S. and A. N. Kolmogorov in the U.S.S.R. Just as most integers are random according to the intuitive meaning of the word, most integers are incompressible or nearly so, because there are far too few short programs to go around. In other words, even if no programs were wasted (say by computing the same result as other programs), only a small fraction of the n-bit integers could be provided with programs even a few bits shorter than themselves.

Using this definition of randomness, Chaitin demonstrated the surprising fact that although most integers are random, only finitely many of them can be proved random within a given consistent axiomatic system. A form of Gödel's famous incompleteness theorem, this result implies in particular that in a system whose axioms and rules of inference can be described in n bits it is not possible to prove the randomness of any integer much longer than n bits. Chaitin's proof of this assertion makes use of a computerized version of the Berry paradox: Suppose in a proof system describable in a small number of bits the randomness of some integer with a large number of bits can be proved. One could then design a small Turing-machine program based on the proof system that would yield the large random integer as output. If the large integer were truly random, however, it could not be the output of any small program, and so a contradiction has been reached.

More precisely, the Turing-machine program would incorporate a fixed supervisory routine, or subprogram, with a length of, say, c bits. This routine would utilize the n bits of axioms and rules of inference to begin systematically generating all possible proofs that could be derived from the axioms in order of increasing number of deductive steps: first all one-step proofs, then all two-step proofs and so on. After each proof was generated the routine would check to see if it was a proof that some integer of considerably more than $n+c$ bits is random. If such a proof were found, the supervisory routine would print out the large random integer specified in the proof and then call a halt to the entire computation. The total length of the Turing-machine program (the supervisory routine plus the axioms and the rules of inference) would be $n+c$ bits, however. In other words, a program $n+c$ bits long would have generated as its output a specific integer that by the algorithmic definition of randomness could not be produced by any program as small as $n+c$ bits. The only escape from this contradiction is to conclude either that the axiomatic system is inconsistent (that is, untrue statements can be proved within it) or that the systematic generation of proofs must continue indefinitely, without uncovering a proof of randomness for any integer much larger than $n+c$ bits. The original Berry paradox appeared as a nuisance, casting doubt on the seemingly meaningful notion of a random integer. The computerized Berry paradox surrounds this notion with a necessary hedge of unprovability, allowing it to be defined in a noncontradictory manner.

Long before the notion of a random integer was taken seriously Émile Borel, Richard von Mises and other mathematicians sought to define and find examples of random real numbers or, equivalently, random infinite sequences of decimal or binary digits. It has been conjectured that irrational numbers such as pi, e and $\sqrt{2}$, which occur naturally in mathematics, are random in the sense that each digit, and indeed each block of digits of fixed length, appears with equal frequency in their decimal expansion. Any sequence of digits with this property is said to be normal. It is not hard to show that no rational number is normal no matter what base it is expressed in, and that almost all the irrational numbers must be normal in every base. So far none of the individual classic irrational numbers has been proved normal, however, although statistical evidence generally supports the conjecture that they are.

On the other hand, it is easy to construct "artificial" irrational numbers that can be proved normal in spite of the fact that their digits follow a trivial and transparent pattern. The most famous of these numbers was invented by D. G. Champernowne in the early 1930's: 0.12345678910111213141516171819202122324. . . . This preposterous number, which consists of the decimal integers in increasing order, has been proved not only irrational and normal (in base 10) but also transcendental. (The transcendental numbers are those numbers that are not the roots of ordinary algebraic equations.) In the initial portions of Champernowne's number there are significant departures from normality, but the differences in frequency tend toward zero as the number of digits being considered is increased. It is apparently not known whether the number is normal in all other bases.

Although the sequence of digits in the decimal expansion of pi may be random in the sense of being normal, the sequence is not unpredictable. In other words, a good gambler, betting on the successive digits of pi, would eventually infer the rule for calculating the number and thereafter win every bet. The same is true of Champernowne's number. Is there a sequence so random that no computable gambling strategy, betting on each successive digit at fair odds, could do better than break even? Any number that is random in this strong sense would necessarily be normal in every base. It is a fundamental result of probability theory that in fact almost all real numbers are random in this way, but finding a specific instance of such a number is not easy. Moreover, there is a sense in which no specifically definable real number can be random, since there are uncountably many real numbers (that is, the set of real numbers is too large to be matched up one-for-one with the positive integers) but only countably many definitions. In other words, the mere fact that a real number is definable makes it atypical. In this case, however, the problem is only to find a number that cannot be shown to be atypical by constructive, or computational, means. In particular the number must not be computable from its definition, because if it were, a perfect betting strategy could be devised.

Chaitin's irrational number Ω is, among its other remarkable properties, random in this strong sense. To understand why that is true, however, it is necessary to deal briefly with the classic unsolvable problem of computability theory known as the halting problem: the task of distinguishing computer programs that come to a spontaneous halt from those that run on indefinitely. Leaving aside gross programming errors, which can cause a program to halt or not to halt for trivial reasons, a program halts if it succeeds in doing what it set out to do, when it has computed, say the 99th prime number or the first million digits of pi. Conversely, a program will run on indefinitely if the task it is embarked on is an unending one, such as computing all prime numbers or searching for a planar map that cannot be colored with only four colors so that no two adjacent regions are the same color.

At first the halting problem might seem solvable. After all, the fact that a program halts can always be demonstrated by simply running the program long enough. Moreover, there are many programs that can easily be proved to halt or not to halt without ever running them. (For example, the famous four-color theorem, which states that five colors are never required for the map-coloring task mentioned above, was finally proved in 1976, and that proof guarantees that the map-coloring program will not halt.) The difficulty comes, then, not in solving particular cases of the halting problem but in solving the problem in general. A. M. Turing, the British mathematician who invented the Turing machine, showed that there is no general prescription for deciding how long to run an arbitrary program so that its halting or nonhalting will be revealed. He also showed that there is no consistent system of axioms strong enough to decide the halting of all programs without running them. The unsolvability of the halting problem can be derived from and indeed is equivalent to the more recent result that randomness, in the Chaitin-Kolmogorov sense of incompressibility, is a property that most integers have but cannot be proved to have.

Now imagine that the Turing machine, instead of being given a definite program at the beginning of its computation, is fed a random sequence of bits. That can be accomplished by flipping a coin whenever the machine requests another bit from its input tape and feeding in 1 or 0, depending on whether the coin comes up heads or tails. This procedure raises a curious question: When the procedure is begun, what is the probability that the machine will eventually halt?

The answer is Chaitin's number Ω. Because the value of Ω depends on which universal Turing machine is being used, it is not a single universal constant like pi. For a given machine, however, Ω is a well-defined irrational number between zero and one, with a natural interpretation as that machine's halting probability on a random program. A randomly chosen program is very likely to tell the computer to do something impossible or pointless, so that the machine either stops immediately in an error state or loops endlessly through a small sequence of instructions. For most computers the former behavior predominates, and so because the halting probability is close to 1 the decimal expansion of Ω begins with several consecutive 9's. It can be shown, however, that the digit sequence of Ω soon becomes quite patternless, ultimately defeating any computable gambling strategy, as well as being random in the Chaitin-Kolmogorov sense of incompressibility.

The most remarkable property of Ω is not its randomness, however. After all, it shares that property with the great majority of real numbers. Rather it is the fact that if the first few thousand digits of Ω were known, they would, at least in principle, suffice to decide most of the interesting open questions in mathematics. This property, as well as Ω's immunity to gambling, is due to the compact way Ω encodes solutions to the halting problem.

The most famous unsolved problem in mathematics is probably Fermat's "last theorem," which states that the equation $x^n + y^n = z^n$ has no solution in positive integers when n exceeds 2. Pierre de Fermat made this assertion in a handwritten note in the margin of a book on number theory, adding that he had discovered a truly remarkable proof of it that the margin was not large enough to hold. Fermat died without exhibiting his proof, and three centuries of effort by other mathematicians have yielded neither a proof nor a refutation.

Fermat's last theorem, like most of the famous unproved conjectures in mathematics, is an assertion of nonexistence and therefore could be refuted by a single finite counterexample, namely a set of integers x, y, z and n that solve the equation. Such finitely refutable conjectures are equivalent to the assertion that some computer program that searches systematically for the allegedly nonexistent object will never halt. Another famous finitely refutable proposition is Goldbach's conjecture, which asserts that every even number is the sum of two primes.

Another kind of finitely refutable conjecture that has had an important place in the history of mathematics is the assertion that some proposition is independent of a given set of axioms, that is, the proposition can be neither proved nor refuted. The most famous propositions of this type are the parallel postulate, which states that through a given point there is exactly one line parallel to a given line, and the continuum hypothesis, which states that there is no infinite number between aleph-null (the number of positive integers) and c (the number of real numbers). In the 19th century the parallel postulate was shown to be independent of the other axioms of Euclidean geometry, and in this century the continuum hypothesis was shown to be independent of the axioms of set theory. The independence of a proposition P from a given set of axioms is equivalent to the nonhalting of a program that systematically generates proofs from the axioms, searching for a proof or refutation of P.

Not all famous conjectures are finitely refutable. For example, no finite amount of direct evidence can decide whether pi is normal, whether there are infinitely many twin primes (consecutive odd primes such as 11 and 13 or 857 and 859) or whether the $P \neq NP$ conjecture in complexity theory is true. [The $P \neq NP$ conjecture asserts that there are mathematical problems for which the validity of guessed solutions can be verified quickly but for which solutions cannot be found quickly.] Such conjectures are not equivalent to halting problems, but there is good reason to believe most of them could be settled indirectly, by proving stronger, finitely refutable conjectures. For example, many twin primes are known, and empirical evidence indicates that the spacing between them grows rather slowly. Therefore the twin-prime conjecture may be viewed as an unnecessarily weak form of a stronger and still probably true assertion about the spacing of twin primes, say that there is always at least one pair of twin primes between 10^n and 10^{n+1}. This stronger assertion is equivalent to the nonhalting of a program that looks for an excessively large gap (greater than a factor of 10) in the distribution of twin primes. (It is important to note that some mathematical questions cannot be reduced to halting problems, for example, some questions about Ω itself. These irreducible questions tend, however, to be rather artificial and self-referent.)

Interesting conjectures, like interesting numbers, tend to be concisely describable. It is hard to imagine a mathematically interesting, finitely refutable conjecture so verbose that it could not be encoded in the halting problem for a small program, one a few thousand or tens of thousands of bits long. Thus the answers to all interesting conjectures of this kind, including those that have yet to be formulated, would in principle be available if there were some kind of "oracle" capable of solving the halting problem for all programs shorter than a few thousand bits. The number of programs involved would still be enormous; for example, there are about $2^{1,000}$ programs shorter than 1,000 bits. Hence it would seem that any oracle able to answer all these questions correctly would have to either be very smart or possess an enormous amount of stored information. In fact, because of the compact way in which Ω encodes the halting problem, its first few thousand bits serves as just such an oracle.

How can solutions to specific halting problems be recovered from Ω? Since Ω is defined as the overall halting probability of a computer with random input, it can be regarded as the sum of the probabilities of all halting computations. Each halting program contributes to the sum its own probability of being chosen (by accident, as it were) when the input bits are supplied by coin tossing. The probability of generating any particular k-bit program in k coin tosses is $1/2^k$. Therefore if feeding this program to the Turing machine one bit at a time causes it to embark on a halting computation in which all k bits of the program but no more are actually requested and read, then that program's contribution to Ω is $1/2^k$. (Programs that call for the machine to read more or fewer bits than there are in the program are considered not to halt, ensuring that each halting computation's contribution to Ω is counted only once.)

Figure 132 shows how the first n bits of Ω can be used to solve the halting problem for all programs of n bits or fewer in length. Because Ω is an irrational number the number that consists of its first n bits, Ω_n, slightly underestimates its true value: Ω_n is less than Ω, which is less than $\Omega_n + 1/2^n$. In order to solve the halting problem for all n-bit programs, one begins a systematic but unending search for all the programs that

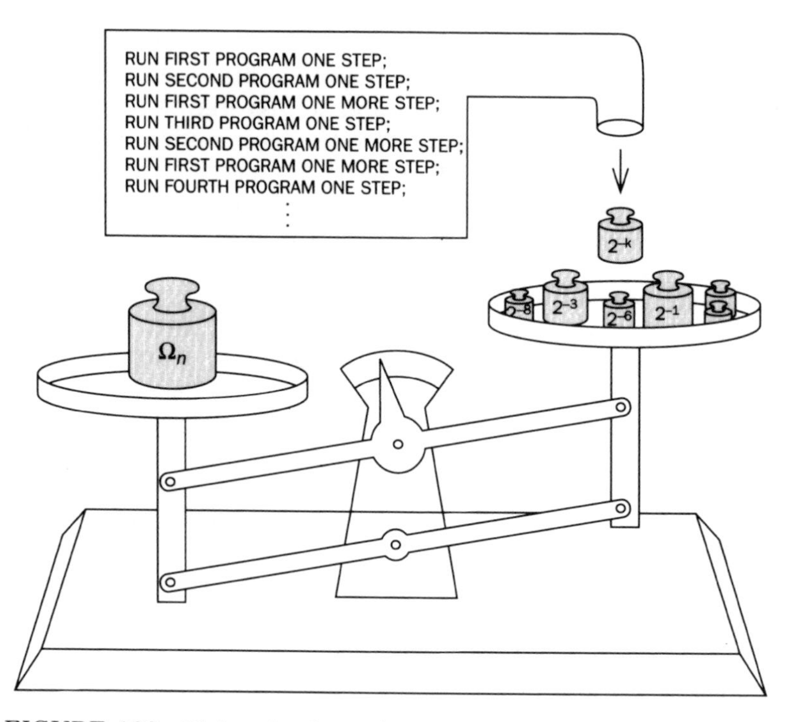

FIGURE 132 Using the first n bits of Ω to solve the halting problem for all programs of n bits or fewer

halt, of whatever length, running first one program and then another for increasingly long times until enough halting programs have been found to account for more than Ω_n in total halting probability.

One way to visualize this process is to consider a balance with a weight equal to Ω_n in its left pan. As is shown in the illustration the programs are run in a way that is reminiscent of the song "The Twelve Days of Christmas": first one step of the first program is run, then one step of the second program and another step of the first, then one step of the third program, another step of the second and another step of the first, and so on. Every time a program of length k is found to halt, a weight of $1/2^k$ is dropped into the right pan of the balance, because $1/2^k$ is the probability of that program being chosen and executed by a computer whose input bits are supplied by coin tossing. Eventually the balance must tip to the right, since the total weight of the programs that halt—the halting probability of Ω—is an irrational number between Ω_n and $\Omega_n + 1/2^n$. By this time a great many programs will have been found to halt, some longer than n bits and some shorter, and many programs that may halt later will not have done so yet. After the balance has tipped, however, no more programs of length n or less can halt, because if one did, that would raise Ω above its established upper bound of $\Omega_n + 1/2^n$. In other words, the subsequent halting of a program of n bits or fewer would alter one of the known digits of Ω.

If this gargantuan computation were carried out with a sufficiently precise estimate of Ω, say the first 5,000 bits, then among the programs whose fate would be decided would be one whose nonhalting would verify Fermat's last theorem, and programs that would decide Goldbach's conjecture and all other simply stated, finitely refutable conjectures. In addition programs would be included whose nonhalting would almost certainly settle many conjectures that are not finitely refutable, such as those about the normality of pi, the twin primes and the $P \neq NP$ question, by proving stronger, finitely refutable statements.

Returning to the senses in which Ω itself is random—its incompressibility and the impossibility of successfully gambling on its digits—it may seem strange that Ω can contain so much information about the halting problem and yet be computationally indistinguishable from a meaningless random sequence. Actually Ω is a totally informative message, one that appears to be random because all redundancy has been squeezed out of it, a message consisting only of information that can be obtained no other way.

To put Ω's lack of redundancy in perspective, consider a more traditional way of encoding the halting problem in an uncomputable irrational number: let K be defined as the real number whose nth bit is 1 or 0, depending on whether or not the nth program halts. K is indeed often referred to as an oracle for the halting problem, but it is a very dilute oracle in the sense that the first 2^n bits of K contain about the same information as the first n bits of Ω: enough to solve the halting problem for all programs of n bits or fewer. K is susceptible to gambling precisely because it is dilute. For example, a sizable portion of all programs can easily be proved to halt or not to halt for trivial reasons. The corresponding bits of K are predictable, and by betting only on those bits a gambler could win consistently. Moreover, even the unpredictable bits of K

are not totally unpredictable. Often two programs can be shown to be attacking the same nontrivial problem in different ways. The halting of one program will then decide the halting of the other, and a gambler who "passed" on the first program, not knowing which way to bet, could bet confidently on the second.

The fact that Ω is incompressible and immune to gambling follows from its compact encoding of the halting problem. Because the first n bits of Ω solve the halting problem for all programs of n bits or fewer, they constitute an "axiom" sufficient to prove the incompressibility of all incompressible integers of n bits or fewer. If Ω_n could be computed by a program significantly shorter than n bits, then a program of similar size would suffice to find and print out the first incompressible n-bit integer, which is a contradiction. In other words, since Ω_n provides enough information to compute a specific n-bit incompressible integer, it must be incompressible itself.

Throughout history mystics and philosophers have sought a compact key to universal wisdom, a finite formula or text that would provide the answer to every question. The use of the Bible, the Koran and the I Ching for divination and the tradition of the secret books of Hermes Trismegistus and the medieval Jewish Cabala exemplify this belief or hope. Such sources of universal wisdom are traditionally protected from casual use by being difficult to find as well as difficult to understand and dangerous to use, tending to answer more questions and deeper ones than the searcher wishes to ask. The esoteric book is, like God, simple but undescribable. It is omniscient, and it transforms all who know it. The use of classical texts to foretell mundane events is considered superstition nowadays, yet in another sense science is in quest of its own Cabala, a concise set of natural laws that would explain all phenomena. In mathematics, where no set of axioms can hope to prove all true statements, the goal might be a concise axiomatization of all "interesting" true statements.

Ω is in many senses a Cabalistic number. It can be known of through human reason, but not known. To know it in detail one must accept its uncomputable sequence of digits on faith, like words of a sacred text. The number embodies an enormous amount of wisdom in a very small space inasmuch as its first few thousand digits, which could be written on a small piece of paper, contain the answers to more mathematical questions than could be written down in the entire universe—

among them all interesting finitely refutable conjectures. The wisdom of Ω is useless precisely because it is universal: the only known way of extracting the solution to one halting problem, say the Fermat conjecture, from Ω is by embarking on a vast computation that would at the same time yield solutions to all other simply stated halting problems, a computation far too large to be actually carried out. Ironically, however, although Ω cannot be computed, it might be generated accidentally by a random process, such as a series of coin tosses or an avalanche that left its digits spelled out in the pattern of boulders on a mountainside. The first few digits of Ω are probably already recorded somewhere in the universe. No mortal discoverer of this treasure, however, could verify its authenticity or make practical use of it.

As this column was going to press I received a telegram from the notorious numerologist Dr. Matrix asserting that he is in possession of the first 1,101,121 bits of Ω (in principle enough to answer a good many uninteresting questions as well as all the interesting ones). He is currently soliciting bids on individual bits or consecutive blocks.

BIBLIOGRAPHY

"Randomness and Mathematical Proof." Gregory J. Chaitin, in *Scientific American,* May, 1975, pages 47–51.

Algorithmic Information Theory. Gregory J. Chaitin. Cambridge University Press, 1987.

Information, Randomness, and Incompleteness. Gregory J. Chaitin. World Scientific, 1987.

"The Ultimate in Undecidability." Ian Stewart, in *Nature,* 232, 1988, pages 115–116.

"Randomness in Arithmetic." Gregory J. Chaitin, in *Scientific American,* July, 1988, pages 79–85.

Name Index